T0225062

SpringerBriefs in Applied Sciences and Technology

SpringerBriefs present concise summaries of cutting-edge research and practical applications across a wide spectrum of fields. Featuring compact volumes of 50 to 125 pages, the series covers a range of content from professional to academic.

Typical publications can be:

- A timely report of state-of-the art methods
- An introduction to or a manual for the application of mathematical or computer techniques
- A bridge between new research results, as published in journal articles
- A snapshot of a hot or emerging topic
- An in-depth case study
- A presentation of core concepts that students must understand in order to make independent contributions

SpringerBriefs are characterized by fast, global electronic dissemination, standard publishing contracts, standardized manuscript preparation and formatting guidelines, and expedited production schedules.

On the one hand, **SpringerBriefs in Applied Sciences and Technology** are devoted to the publication of fundamentals and applications within the different classical engineering disciplines as well as in interdisciplinary fields that recently emerged between these areas. On the other hand, as the boundary separating fundamental research and applied technology is more and more dissolving, this series is particularly open to trans-disciplinary topics between fundamental science and engineering.

Indexed by EI-Compendex, SCOPUS and Springerlink.

More information about this series at http://www.springer.com/series/8884

Lumin Chen · Yihao Li · Lina Han ·
Liang Yuan · Yuxiang Sun · Xiaolu Tang

Elderly Health Services and Remote Health Monitoring

Springer

Lumin Chen
College of Mechanical and Electrical
Engineering
Zhengzhou University of Light Industry
Zhengzhou, China

Yihao Li
College of Mechanical and Electrical
Engineering
Zhengzhou University of Light Industry
Zhengzhou, China

Lina Han
Department of Cardiovascular Internal
Medicine
Chinese PLA General Hospital
Beijing, China

Liang Yuan
College of Mechanical and Electrical
Engineering
Zhengzhou University of Light Industry
Zhengzhou, China

Yuxiang Sun
College of Mechanical and Electrical
Engineering
Zhengzhou University of Light Industry
Zhengzhou, China

Xiaolu Tang
Luoyang Bearing Research Institute Co.,
Ltd.
Luoyang, China

ISSN 2191-530X   ISSN 2191-5318   (electronic)
SpringerBriefs in Applied Sciences and Technology
ISBN 978-981-15-7153-4   ISBN 978-981-15-7154-1   (eBook)
https://doi.org/10.1007/978-981-15-7154-1

This Springer imprint is published by the registered company Springer Nature Singapore Pte Ltd.
The registered company address is: 152 Beach Road, #21-01/04 Gateway East, Singapore 189721,
Singapore

Contents

Contributors

Jizhao Gao College of Mechanical and Electrical Engineering, Zhengzhou University of Light Industry, Zhengzhou, Henan Province, P.R. of China

Sufeng Guo College of Mechanical and Electrical Engineering, Zhengzhou University of Light Industry, Zhengzhou, Henan Province, P.R. of China

Yi Gao College of Mechanical and Electrical Engineering, Zhengzhou University of Light Industry, Zhengzhou, Henan Province, P.R. of China

Lina Han Department of Cardiovascular Internal Medicine, Second Medical Center, Chinese PLA General Hospital, Beijing, China

Zhuangyu Hu College of Mechanical and Electrical Engineering, Zhengzhou University of Light Industry, Zhengzhou, Henan Province, P.R. of China

Guohao Huang College of Mechanical and Electrical Engineering, Zhengzhou University of Light Industry, Zhengzhou, Henan Province, P.R. of China

Junying Li College of Mechanical and Electrical Engineering, Zhengzhou University of Light Industry, Zhengzhou, Henan Province, P.R. of China

Aodong Li College of Mechanical and Electrical Engineering, Zhengzhou University of Light Industry, Zhengzhou, Henan Province, P.R. of China

Shixiang Li College of Mechanical and Electrical Engineering, Zhengzhou University of Light Industry, Zhengzhou, Henan Province, P.R. of China

Kejie Li College of Mechanical and Electrical Engineering, Zhengzhou University of Light Industry, Zhengzhou, Henan Province, P.R. of China

Yihao Li College of Mechanical and Electrical Engineering, Zhengzhou University of Light Industry, Zhengzhou, Henan Province, P.R. of China

Ruxin Liang College of Mechanical and Electrical Engineering, Zhengzhou University of Light Industry, Zhengzhou, Henan Province, P.R. of China

Xiangye Liu College of Mechanical and Electrical Engineering, Zhengzhou University of Light Industry, Zhengzhou, Henan Province, P.R. of China

Jin Ma College of Arts and Design, Zhengzhou University of Light Industry, Zhengzhou, Henan Province, P.R. of China

Hubiao Tang College of Mechanical and Electrical Engineering, Zhengzhou University of Light Industry, Zhengzhou, Henan Province, P.R. of China

Sai Wang College of Mechanical and Electrical Engineering, Zhengzhou University of Light Industry, Zhengzhou, Henan Province, P.R. of China

Hao Wu College of Mechanical and Electrical Engineering, Zhengzhou University of Light Industry, Zhengzhou, Henan Province, P.R. of China

Xiaolu Tang Luoyang Bearing Research Institute Co., Ltd., Luoyang, Henan Province, P.R. of China

Hu Xu College of Mechanical and Electrical Engineering, Zhengzhou University of Light Industry, Zhengzhou, Henan Province, P.R. of China

Yong Xu College of Mechanical and Electrical Engineering, Zhengzhou University of Light Industry, Zhengzhou, Henan Province, P.R. of China

Zuyu Yao College of Mechanical and Electrical Engineering, Zhengzhou University of Light Industry, Zhengzhou, Henan Province, P.R. of China

Liang Yuan College of Mechanical and Electrical Engineering, Zhengzhou University of Light Industry, Zhengzhou, Henan Province, P.R. of China

Lei Yue College of Mechanical and Electrical Engineering, Zhengzhou University of Light Industry, Zhengzhou, Henan Province, P.R. of China

Sun Yuxiang College of Mechanical and Electrical Engineering, Zhengzhou University of Light Industry, Zhengzhou, Henan Province, P.R. of China

Qi Zhang College of Mechanical and Electrical Engineering, Zhengzhou University of Light Industry, Zhengzhou, Henan Province, P.R. of China

Qian Zhang College of Mechanical and Electrical Engineering, Zhengzhou University of Light Industry, Zhengzhou, Henan Province, P.R. of China

Yuzhu Zheng College of Mechanical and Electrical Engineering, Zhengzhou University of Light Industry, Zhengzhou, Henan Province, P.R. of China

Chapter 1
Simulation and Experimental Analysis of Auxiliary Lower Extremity Exoskeleton

1.1 Introduction

As the proportion of the elderly in society is increasing rapidly, the number of people over 65 will increase from 17.4 to 29.5% of the total population between 2010 and 2060 [1]. In order to improve their quality of life, social participation and physical support for the elderly are necessary. In addition, for elderly people with functional diseases such as neuromuscular diseases, active daily life can also be promoted through appropriate physical support. Today, wearable devices are expected to be one of the tangible solutions to these aging social problems, as they are used to assist groups with low muscle strength.

Since human joints have complex motions and cannot be simulated as rotatory joints, the exact position of human joints cannot be determined externally form the outside without imaging devices [2], the resulting misalignment of the joint axis can be uncomfortable and may cause pain or even long-term injury during repeated use. Domestic and foreign teams have proposed several exoskeleton robots that can perform lower limbs, especially complex joint movements of the knee. Celebi et al. introduced a self-aligning active exoskeleton for robot-assisted knee rehabilitation [3]. The authors utilized a planar parallel kinematic chain, commonly referred to as a Schmidt coupling. Moreover, the knee joint featured an active rotational 1-DoF controlled series elastic actuator and two passive translational DoF in the sagittal plane. Stienen et al. proposed a self aligning elbow joint mechanism with two extra links to support the human elbow joint [4]. The translation of the joint is independent of rotation. Amigo et al. proposed a 3-DoF joint system with dynamic servo-adaptation. Their knee joint has three active DoF in the sagittal plane.

The paper mainly designs a kind of rope-type lower limb auxiliary device by modeling the human joint muscle and tendon driving, studies the kinematics and dynamics of the device. The cable module in ADAMS simulates the rope drive and judges the rationality of the structure. At the same time, the range test and pressure analysis are performed on the physical model to ensure that the elderly are not injured when they walk.

1.2 Design of Lower an Extremity Exoskeleton Structure

This section mainly in reference biological structure characteristics of the human knee ankle after reasonable on knee ankle structured to design.

1.2.1 Design of a Knee Joint

The human knee joint is one of the most complex joints in the human body. The human knee connects the thigh bone to the tibia. The smaller bones alongside the tibia and kneecap are the other bones that make up the knee joint. The human knee joint can be kinematics modeling as a 6-DoF joint. However, flexion/extension motion is the dominant movement in the sagittal plane of the knee because several-degree-of-freedom motions are limited to the interaction of strong ligaments and muscles. The condyle of the femur is shaped like an oval, while the upper tibia is shaped like a plane. Thus, when the knee is bent for flexion/extension, the femur rolls relative to the tibia, causing forward and backward translation. As above, the rolling and sliding motion between the femur and tibia results in significant forward and backward translation over 19 mm in the sagittal plane [5]. Moreover, the femur rolls relative to the tibia and rotates simultaneously while the knee is bent. Due to this coupling of translational and rotational motion, the center of rotation moves with the bending angle of the knee. The instantaneous center of rotation is at the intersection of the long axis of the tibia and femur, as shown in Fig. 1.1.

The knee joint of a general wearable robot is implemented as a simple rotary joint for pure rotation in the sagittal plane. Therefore, even if their initial positions are carefully aligned, the center of rotation of the knee joint of the person and the wearable device for the lower limb assist will not be aligned with the bending of the knee [6]. We have proposed a movable knee joint mechanism that can be automatically aligned to assist in the flexion/stretching of the knee joint, as shown in Fig. 1.2. In order to compensate for the motion of the instantaneous center of rotation by the flexion/extension motion of the human knee joint, a joint mechanism consisting of two passive degrees of freedom and one active degree of freedom is proposed. The self-aligning knee joint, achieved by adding two redundant degrees of freedom on an active rotating joint, that compensates for the misalignment in the center of rotation between the wearer's knee joint and the joint of the wearable robot. The automatic alignment mechanism in the knee joint consists of two linkage and four pulleys, compared with Stienen's research, the institution can support part of the patient's weight, as shown in Fig. 1.3.

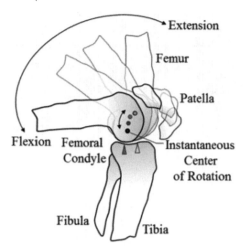

Fig. 1.1 Buckling/stretching of the knee joint in the sagittal plane

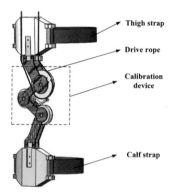

Fig. 1.2 Schematic diagram of the knee joint

Research, the institution can support part of the patient's weight, as shown in Fig. 1.3.

The schematic diagram of the automatic calibration device is shown in Fig. 1.2. The driving pulley and the aligning pulley 1 are coaxially fixed and can transmit the same torque. The aligning pulley 1 and the aligning pulley 2 are connected through the connecting rod 1. The aligning pulley 2 and the aligning pulley 3 pass through. The connecting rod 2 is connected, and the aligning pulley 1, the aligning pulley 2, and the aligning pulley 3 transmit the same torque in sequence through the calibration rope. Therefore, the aligning pulley 3 and the driving pulley have the same torque; the aligning pulley 3 is connected to the lower limb and the lower leg. The bars are fixedly connected and provide the same torque. This device can be regarded as composed of two parallelograms in series. The parallelogram structure is assembled

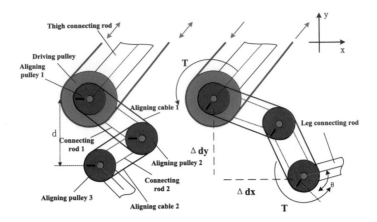

Fig. 1.3 Three-dimensional model of knee exoskeleton

by a calibration rope and a connecting rod. This mechanism can transfer the torque from the thigh to the calf, and make the joint rotation and the instantaneous center rotation axis. Irrelevant, this can lift the strict restriction of the human-machine joint on the heart. The knee exoskeleton is mainly divided into two states when assisting the human body in rehabilitation training: standing and knee flexion; when the knee exoskeleton assists the human body to stand, it acts as a standing support for the human body. The line contact method plays a supporting role. The connecting rod 1 and the aligning pulley 2 in the automatic calibration device act on the lower leg connecting frame, which plays a supporting role. When assisting the human body to bend the knee, the automatic calibration device plays a key role. The connecting rod 1 and the connecting rod 2 can rotate at the same time, adjust the position of the rotation center and transmit the rotation angle, thereby promoting the exoskeleton to assist the human body to perform knee flexion training while changing the instantaneous center of movement.

The mechanism consists of three small pulleys of the same diameter connected together, with two links (l_1, l_2) and passive operation. With this, each angle can be set to any value to compensate for misalignment. In order to detect the angle, the pulley is equipped with an angle sensor. In this case, the translation workspace (x_2, y_2) used for compensation is calculated as a function of the angle of the measurements, as follows:

$$x_2 = l_1 \cos \theta_1 + l_2 \cos(\theta_1 + \theta_2) \tag{1.1}$$

$$y_2 = l_1 \sin \theta_1 + l_2 \sin(\theta_1 + \theta_2) \tag{1.2}$$

1.2.2 Design of an Ankle Joint

During the standing phase, when the foot is in contact with the ground, kinematic dissimilarity at the ankle can cause accidental movement deviations in all the proximal frames and joints of the device. In order for the device to fit tightly under the wearer, the relative movement between the wearer and the device should be minimized.

The ankle joint design should mainly consider the flexibility of movement, the rigidity of load and gravity transmission, and the anthropomorphic characteristics of man-machine integration. The flexion and extension freedom and left-right swing are designed at the ankle. To ensure that the wearer can turn normally, left-right rotation must be added at the ankle. When the wearer walks normally, the ankle joint is accompanied by flexion and extension and left-right swing. Therefore, the design of the ankle needs to add three degrees of freedom of flexion and extension, left-right swing, and left-right rotation. In order to meet the lightweight design requirements and meet the anthropomorphic design principles, bearings joint is used at the ankle joint of the exoskeleton to achieve 3 degrees of freedom of movement [7]. There is a casing sleeve between the bearing joint and the foot plate. There are 3 springs in the sleeve to keep the length of the sleeve small during the movement. The foot plate is fixedly connected to the cylindrical sleeve under the ankle joint, the bearing joint rod, the spring and the upper seat of the pressure sensor are connected, and the upper end can be moved against the pressure sensor. A signal is output when the force is applied, so that the pressure value curve formed by each movement can be monitored in real time. If the maximum value is within a reasonable range, it will not cause harm to the human body. The specific structure is shown in Fig. 1.4.

The exoskeleton foot sole consists of a cushioned rubber pad, a contact detection module, and a patient's shoe. The soles of the feet are mostly made of rubber or

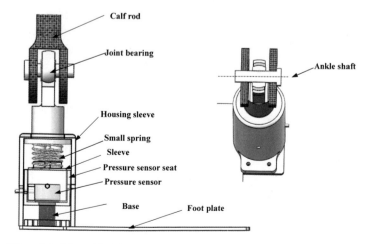

Fig. 1.4 3D model of the ankle joint

silicone, which guarantees a certain degree of flexibility and can follow the wearer's corresponding bending when exercising. The bottom of the foot is tied to the patient's shoes to meet the needs of different patients. At the same time, a flexible button contact sensor needs to be placed in the shoes. Its sensitivity is high, and it can detect the human feet and exoskeleton feet when the wearer just lifts his feet. The signal of the detachment of the head and the contact signal of the exoskeleton's feet and the ground when the exoskeleton is about to contact the ground effectively improves the system safety and rapid response.

1.3 Simulation Analysis Based on ADAMS Cable Module

1.3.1 ADAMS Cable Modeling Principle and Model Simulation

Cable is a new plug-in that was added after the ADAMS 2013 version. The purpose is to help users quickly build a rope simulation model and complete the engineering application. Compared with the previous model using the sleeve, the Cable module features Anchor, Pully and Roller modeling capabilities for fast parametric modeling [8].

The model is established in ADAMS_cable, where the geometry and contact parameters of the pulley are shown in Table 1.1. The gravity direction is set as the negative direction of the Y-axis, the pulley material is set as steel, and the specific parameters of the rope are shown in Table 1.2. Firstly, the change of the length of the rope on both sides is obtained as the driving function, and the joint angle is given as the angle change of the gait cycle of the knee joint of the human lower limb [9].

The knee joint pulley-rope model is constructed as show in Fig. 1.5. In the figure, the coordinates of the pulley center are pulley 1, 2 (0, 0, −4); pulley 3 (−52, −30, −4); pulley 4 (−5, −65, −4), the pulley 1 and the pulley 2 is concentric and fixed, and the two ends of the rope 2 are fixed on the slide rails of the pulley 2 and the

Table 1.1 Pulley geometry and contact parameter settings

Parameter name	Pulley1	Pulley2	Pulley 3	Pulley4
Pulley width/mm	4	4	4	4
Pulley depth/mm	2	2	2	2
Pulley arc radius/mm	1	1	1	1
Angle/°	20	20	20	20
Pulley radius/mm	30	15	15	15
Contact stiffness coefficient/N/mm	$1*10^5$	$1*10^5$	$1*10^5$	$1*10^5$
Friction factor	0.6	0.6	0.6	0.6
Contact point normal relative speed/mm/s	100	100	100	100

Table 1.2 Rope parameter setting

Parameter name	Rope 1	Rope 2	Rope 3
Elastic modulus/Pa	1.0E+004	1.0E+004	1.0E+004
Damping coefficient/(N/mm*s)	1.0E−002	1.0E−002	1.0E−002
Rope diameter/mm	2	2	2

Fig. 1.5 Knee joint simulation model

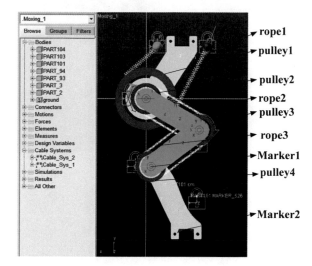

pulley 4, so that the pulleys 1, 2, 3 are sequentially connected. Both ends of the rope 3 are fixed to the slide rails of the pulley 2 and the pulley 4, so that the pulleys 3, 2, 1 are sequentially connected. The greater the elastic modulus of the rope material, the smaller the elastic deformation of the rope under the same force, the greater the stiffness, the less deformed the material.

1.3.2 Simulation Analysis of 'Walking' Model

Gait analysis is an indispensable technical link for exoskeleton robots [10]. Set the simulation time to 4 s and the step size to 500 steps. When the massless "red ball" in Fig. 1.4 is pulled in the direction of the rope for 2 s, the knee joint reaches the maximum bending angle when walking. After 2 s, when the mass "green ball" is pulled by 2 s in the direction of the rope, the human leg returns to the vertical support position. That is, one cycle is 4 s, and the simulation process is shown in Fig. 1.6.

Fig. 1.6 The process of knee joint rotation for one cycle

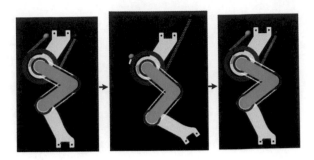

Fig. 1.7 Angle curve

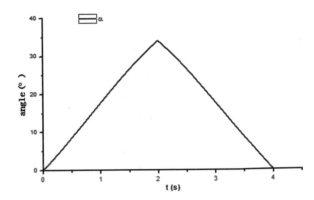

1.3.2.1 Simulation Corner Analysis

The simulation results are shown in Fig. 1.7. In the case of walking, the angular change of the joint measurement is an acute angle between the abstract thigh rod and the calf rod. In a period of 4 s, when the time is 2 s, the angle reaches the peak value 35°. In order to verify the correctness of the rope drive, it is necessary to compare the simulated joint rotation angle with the actual joint rotation angle [11], which is in accordance with the angle of rotation of the knee joint when the person walks.

In addition, the joint is capable of performing up to 145° rotation in the sagittal plane. Therefore, the anthropomorphic knee joint can effectively reduce the wearer's resistance during normal exercise and provide sufficient space for the wearer to assist the gait.

1.3.2.2 Simulation Trajectory Analysis

In the simulation process, the displacement of the knee joint is measured by trajectory tracking of Marker1 and Marker2 during flexion and extension. As shown in Fig. 1.8, the joint point Marker1 is performing a flexion and extension movement with a half period of 2 s, that is, when the knee joint is flexed to 35°, the translation distance in the vertical direction Y can be up to 20 mm. As shown in Fig. 1.9, the joint point

Fig. 1.8 Marker1's
trajectory

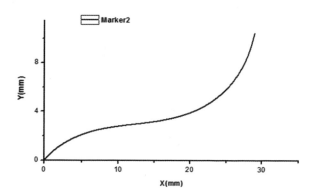

Fig. 1.9 Marker2's
trajectory

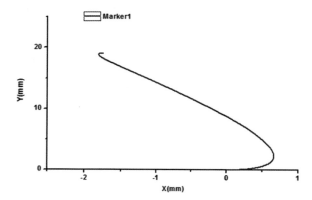

Marker2 is in the trajectory of the flexion and extension movement of half a cycle (i.e., 2 s), forming a two-dimensional translational working space of about 30 mm × 12 mm.

Combining the motion trajectories of Marker1 and Marker2, the proposed joint mechanism has a two-dimensional translational working space of about 33 mm × 32 mm without the separation of the rods.

1.4 Experiment Analysis

An experimental assessment of the performance of the knee joint and the auxiliary platform is described. The first experiment tested the basic performance of the knee joint, especially in terms of range of motion and maximum thickness. The second experiment examined the auxiliary performance of the knee joint for joint pressure and translational workspace.

1.4.1 Basic Performance Test

The knee joint is capable of performing a 145° rotation in the sagittal plane. In order to verify that it is free to move, we have tried various postures encountered in the daily life of the wearer. As shown in Fig. 1.10a, the wearer does not feel the gravity of the object when he poses as a "standing" position, thereby determining that it satisfies our goal. As shown in Fig. 1.10b, the wearer puts a "sit down" posture, and the knee joint and the elastic ankle joint automatically adjusts the angle, which is reflected in the change of the distance from the sole to the ankle axis when the ankle is rotated. As shown in Fig. 1.10c, the knee joint can be adjusted to an angle of 145° when the wearer poses in the "squat" position. The pressure sensor at the ankle joint can display the external pressure on the ankle joint in different motion postures in real time. In addition, the weight support structure can support part of the wearer's weight and platform load by transmitting force to the ground.

Table 1.3 gives the characterization results of the knee joint mechanism. The maximum thickness of the adjustment structure of the knee joint measured by the coronal method was 28.5 mm. These basic performance tests results show that the knee joint can be tightly fitted to the wearer's lower limbs under vertical load, and can

(a)"Standing"Position   (b)"Sitdown"Position   (c)"Squat" Position

Fig. 1.10 Test range of motion

Table 1.3 Specifications of knee parameters	Angle of rotation	0°–145°
	Weight	400 g
	Maximum thickness	28.5 mm
	The straps exert maximum pressure on the human leg	6 N
	Maximum assist torque	20 NM
	Workspace (Marker 1, 2)	0 mm $\leq$ X $\leq$ 33 mm; 0 mm $\leq$ Y $\leq$ 32 mm

adapt to the daily exercise of the individual even without any resistance or pressure on the wearer's knee.

1.4.2 Pressure Test

In order to verify the auxiliary performance of the automatic alignment of the knee joint and the elastic ankle joint, we asked a healthy male with a weight of 65 kg to wear a sitting-up movement with the auxiliary lower extremity exoskeleton. The experimental platform, as shown in Fig. 1.11, connects the cylindrical pressure sensor SBT674 built in the ankle joint to the display SBT961, and is connected to the computer through a USB to rs232 serial converter. In the sitting motion, the process of joint pressure change is measured, as shown in Fig. 1.12. During the sitting process, six consecutive sitting and standing movements were measured. The pressure curve is periodic and the peak pressure of the joint is 16.3 N. This force will not cause

Fig. 1.11 Pressure measurement experimental platform

Fig. 1.12 "Sitting" pressure curve

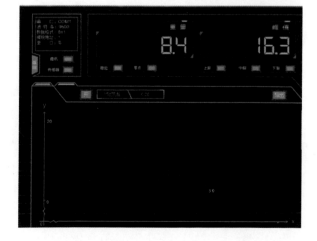

Fig. 1.13 "Walking" pressure curve

damage to the ankle joint [12]. During normal walking, the change in joint pressure is measured. As shown in Fig. 1.13, the pressure curve of 11 steps is measured. It has a periodicity. When the foot is lifted, the pressure value tends to 0. When the foot is stepped on the ground, the pressure value is about 10 N. About "Sit down" and "walk" actions, the pressure detected will not cause discomfort in the ankle of the human body.

1.5 Conclusion

In this article, we introduce an automatic adjustment of the knee joint mechanism and the elastic ankle joint mechanism for the wearable exoskeleton to promote the normal life of the elderly and the weak. The center of rotation of the human knee joint moves with the bending/stretching motion. The self-aligning knee joint compensates for the deviation of the center of rotation between the wearer's knee joint and the joint of the wearable device. The elastic ankle joint can also fine tunes the center of rotation according to the expansion and contraction of the muscle. The system adds redundant degrees of freedom (two DoF totals) to the 1-DoF rotary joint. This has the following advantages: The proposed mechanism can perform joint motion according to the body type without individual deviation. In addition, the joint meets the wearer's desire to feel only minimal resistance during walking and the weight is supported/transmitted to the ground. We conducted a feasibility test. The availability of the proposed self-aligning knee joint mechanism and elastic ankle joint was demonstrated by bending/stretching motion.

The proposed wearable exoskeleton is the first prototype. In order to achieve our goals, there are still some major challenges. To further improve its usability, the safety of the drive system of the wearable exoskeleton should be considered to prevent damage. Additionally, a three-dimensional joint mechanism should be adopted to maximize ergonomic and tightly fitted characteristics. Furthermore, case studies of wearable exoskeletons are needed for older wearers.

Funding Fund Project: National key research and development plan key special project 2017YFF0207400 "Research on key technologies and important standards for health services and remote health monitoring for the elderly and the disabled.

References

1. United Nations Population Division. (2019). *World Population Ageing 2019* [EB/OL]. https://www.un.org/en/development/desa/population/theme/ageing/index.asp.
2. Cenciarini, M., & Dollar, A. M. (2011). Biomechanical considerations in the design of lower limb exoskeletons. In *2011 IEEE international conference on rehabilitation robotics*, Zurich, pp. 1–6.
3. Celebi, B., Yalcin, M., Patoglu, V. (2013). AssistOn-Knee: A self-aligning knee exoskeleton. In *2013 IEEE/RSJ International Conference on Intelligent Robots and Systems*, Tokyo, pp. 996–1002.
4. Stienen, A. H. A., Hekman, E. E. G., van der Helm, F. C. T., et al. (2009). Self-aligning exoskeleton axes through decoupling of joint rotations and translations. *IEEE Transactions on Robotics, 25*(3), 628–633.
5. Bellmans, J., Ries, M. D., & Victor, J. M. K. (2005). *Total knee arthroplasty: A guide to get better performance* (p. 130134). Berlin Heidelberg: Springer.
6. Li, Y. B., Tang, Z. H., & Ji, L. H. et al. (2019). Effects of lower extremity exoskeleton human-machine interconnection devices on joint internal forces. *Journal of Tsinghua University (Science and Technology), 59*(7), 544–550.
7. Hemmings, G. (2018). Joint arthroplasty: Understanding its bioceramics and alternative bearings. *Tritech Digital Media.*
8. Guo, W. D., Li. S. Z., Ma. L. (2015). *ADAMS2013 application example elaboration tutorial*, p. 118. Machinery Industry press, Beijing.
9. Chen, Y. (2016). *Research on walking movement of humanoid robot based on human motion law and foot characteristic*. Hefei: Hefei University of Technology.
10. Wang, X. T., Zhang, J. X., Su, H. L., et al. (2013). Research of lower extremity exoskeleton virtual prototype design. *Mechanical Design & Manufacturing, 5*, 140–142.
11. Zhan, L. Q., Han, Q., Wu, M. H., et al. (2018). Study on knee joint stability during walking in different age groups. *Chinese Journal of Rehabilitation Medicine, 33*(12), 47–50.
12. Tian, C. K., Xu, W. Q. (2005). Research progress in biomechanics of ankle joint flexion and dorsiflexion muscles. *Journal of Beijing Sport University, 28*(11), 1527–1528,1540.

Chapter 2
Generative Design and Simulation of the Exoskeleton in Patients with Scoliosis

2.1 Introduction

Scoliosis is a pathological condition in which the spine is laterally curved in one or more segments. It is a progressive lateral curvature of the spine, often accompanied by vertebral rotation and rib deformation. According to surveys, most scoliosis occurs before puberty, affecting approximately 1.1 to 2.9% of children [1]. According to the etiology, this disease can be divided into two types: idiopathic scoliosis and congenital scoliosis. The condition of scoliosis may bring different degrees of influence on patients. Mild scoliosis patients generally have no obvious discomfort in clinical, and no apparent bony deformity can be seen in appearance; in severe cases, scoliosis can cause back pain, body deformities, and even lung dysfunction. Patients can seek different treatment methods according to their condition. The correction of scoliosis is to make the deformity get the maximum correction and keep it in the position of correction. Treatment methods include non-surgical treatment and surgical treatment. Non-surgical treatment includes electrical stimulation, orthotics, and artificial traction. One of the most effective ways is to wear orthotics. Studies have shown that the effective rate of adolescent idiopathic scoliosis can reach 75% if the patient wears it for a certain period according to the doctor's requirements, and the correction intensity conforms to the predetermined plan [2]. The appliance can provide passive or active force for the treatment of scoliosis so that scoliosis can be corrected to the maximum extent. There are many types of orthotics, including Boston [3] brace, Charleston [4] brace, Lyon [5] brace, Milwaukee [6] brace, etc. According to the current research, it is not sure whether one design is better than the other. However, most of these orthotics are of skeleton structure, the material selection and collocation are unreasonable, the overall weight is massive; the structural design is absurd, the force application joint is not apparent, and it is not convenient for patients to wear; if it is a plaster vest orthotics, there will be a significant error in the production of mold opening, which cannot be adjusted with time, and the utilization rate is low.

Given the above pain points, this paper designed a rehabilitation exoskeleton for scoliosis patients. The exoskeleton described in this paper adopts generative design, with the thinking of bionic, so that the exoskeleton and human body more fit; Accord with ergonomic principle, the plan is more humanized; At the same time, the structure of the exoskeleton has been optimized to reduce its weight. After verification, the exoskeleton can achieve the same effect as ordinary orthotics, but its influence is only 40–60% of regular orthotics. The generated exoskeleton can be used for correction in non-surgical treatment and auxiliary rehabilitation exercise after surgical treatment.

2.2   Methodology

The most commonly used judgment method for scoliosis in clinical practice is the cobb angle (Fig. 2.1) method [7]. The angle formed by the intersection of the perpendicular line of the extension line of the upper margin of the upper-end vertebra and the vertical line of the extension line of the lower margin of the lower end vertebra is cobb angle.

The treatment of scoliosis can be divided into surgical procedures and non-surgical treatment. Generally, surgical treatment is considered when the age is more than ten years old, and the cobb angle is more than 40°. Non-operative treatment is mainly divided into electrical stimulation treatment, orthopedic appliance correction, and

Fig. 2.1 Measurement of scoliosis angle (Cobb)

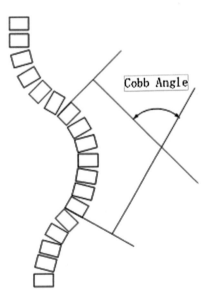

traction treatment. The principle of electrical stimulation is to stimulate the contraction and relaxation of paravertebral muscles, and its treatment effect is not stable; traction treatment needs to be nailed on the bone, which is easy to infect, and this method will make patients feel pain. According to Hueter-Volkmann law: if the pressure on the bone increases, the growth of bone will be inhibited; otherwise, if the pressure on the epiphysis decreases, the growth of bone will be accelerated accordingly [8].

The basic biomechanical principle of brace treatment is: on the coronal plane, the spine is fixed and corrected by a three-point action system. In order to control or improve scoliosis, the load of the growth endplate on the concave side of the scoliosis crest region was reduced, which stimulated the growth of the vertebral body in the concave side of scoliosis, and promoted the reconstruction of the vertebral body structure. At the same time, in the pressure area and release area, the pressure pad is used to apply pressure to cause the displacement of the spine on the coronal plane and the torsion on the horizontal plane, and space is left in the corresponding direction for the release of pressure.

The principle of orthotic appliance correction introduced in this paper (Fig. 2.2), the applied force is provided by the binding strap, its biomechanical principle is based on the most basic three-point mechanical principle, applying a direct and continuous force on lumbar scoliosis, so that the curvature of the spine is transferred from the convex side to the concave side.

Fig. 2.2 Principle of treatment and correction of brace

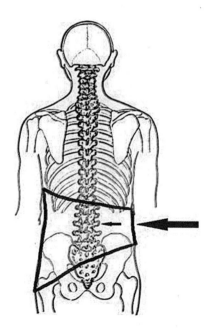

2.3 Design Details

With the development of computer technology, computer aided design (CAD) technology has been widely used in the design of braces. Compared with traditional means, the braces designed by computers are more modern and comfortable [9].

In the design of the brace proposed in this paper, we chose Autodesk Fusion 360, an industrial-grade three-dimensional modeling software that can freely carry out modeling design, which has the advantage that the appearance of the brace cannot be affected by the size of the human body.

If we need to design the brace according to the size of the human body, the first problem we need to solve is to require a three-dimensional social body model. We use the technology of reverse engineering three-dimensional modeling, scan the three-dimensional human body model, and upload it to the modeling software. In this process, we get the three-dimensional social body model, including the data of the bust, waist, and so on, then based on the three-dimensional model of the human body, we designed the brace fitting the human body.

Generative design is a brand-new method and concept. It is mainly to generate different three-dimensional model output results according to the changes of parameters through a set of composition rules or an algorithm. Generative design methods are rooted in system dynamics modeling. Their nature means that they are very suitable for repetitive processes, that is, solutions are obtained through a large number of design operations and iterations.

The generative design process of the exoskeleton is as follows: first, put the unmodified model into the special generative design modeling software; select the elements to be retained in the model, including ribs on both sides of the exoskeleton ribs, back spine groove, shoulder brace and chest guard plate; then, select the obstacle elements, that is, the structure to be optimized in the design. In the cloud computing, the construction of this part chosen will be repeatedly modified; next, the necessary parts will be constrained according to the actual situation of the design, the load will be applied to the parts that need to be stressed, the materials will be given to the brace, and the processing method will be selected. After a period of cloud computing, the software will generate a massive load as required, and the mass is minimal output results.

The most direct effect of the generative design on the brace (Fig. 2.3) is to greatly reduce the weight of the brace, making the brace light and comfortable. In appearance, with a more precious sense of science and technology, which to a certain extent, can slow the boredom and fear of patients for brace.

Based on the complete model, we introduce the concept of generative design. The essence of generative design is to imitate plants in nature, which have evolved over thousands of years to have unique plant roots and branches and leaves. Generative design is a approach of using cloud computing to assign a particular plant structure to its parts. In this paper, we apply this design to the lateral ribs of the exoskeleton. It should be noted that the calculation of generative design is all carried out in the cloud, so final result is uncontrollable. In order to minimize the error rate of the design

Fig. 2.3 The design of
exoskeleton for correction
and rehabilitation of
scoliosis patients

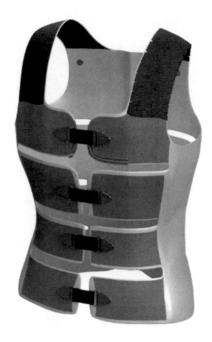

result, the model needs to be simplified, as shown in the Fig. 2.4. After submitting
the model to the cloud according to the steps, after several experiments, the result is
similar to the following figure. In the image below (Fig. 2.4), instead of the traditional
plate-like structure, a generated design replaces it with a root-like structure, which
makes the overall structure of the exoskeleton lighter.

Fig. 2.4 The model after
generative design

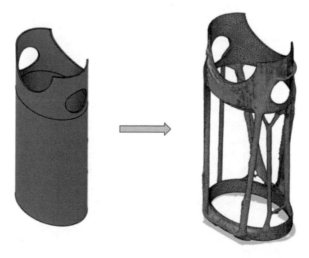

2.4 Simulation and Experiment

After the generative design, the lightweight of the rehabilitation exoskeleton for scoliosis is to achieve better appearance and treatment effect by improving the structure based on satisfying certain medical rehabilitation. Simulations are performed to determine how loads cause deformation and failure, and to understand whether and how parts fail. In this paper, the static analysis and simulation are carried out for the correction of scoliosis exoskeleton, and the deformation and stress in the model are analyzed by structural load and constraint. Through the results, the displacement, stress, and the linear response of the corresponding stress can be studied. After the analysis, we can modify the relevant parts of the model, improve the overall design work, and customize the structural design according to the characteristics and requirements of patients.

Scoliosis correction rehabilitation exoskeleton adjust the degree of tightness through the bandage on the basis of basic medical rehabilitation, drive the soft and hard parts of the exoskeleton body, and make the exoskeleton reach three or four points of stress in the pressure area and release area to correct scoliosis. In this paper, the static simulation analysis is carried out by using the simulation module of Autodesk Fusion 360 software. According to the boundary conditions in the actual wearing process and the load state when the bandage is tensioned, the engineering stress is verified, and the optimized performance state is obtained.

The steps of static analysis and simulation with Autodesk Fusion 360 software for exoskeletons are as follows: (1) select static analysis module to import the model; (2) simplify the model and delete unnecessary detailed features; (3) define model materials; (4) impose constraints to constrain the freedom of the model; (5) add loads; (6) mesh; (7) submit the cloud simulation, view the simulation feedback results and analyze them.

The static analysis and simulation of the rehabilitation exoskeleton for scoliosis correction were carried out. First, we did static analysis and simulation for the model after generative design and the model without generative design, and then compared the results of stress and deformation analysis under the two forms, and finally got the mechanical performance index.

The model without generative design is introduced into the static analysis simulation. The model is simplified, and some features that are not necessary for simulation are deleted. After simplification, it is necessary to define the material for the exoskeleton. Since the exoskeleton is designed to combine the hard and soft parts inside and outside, only the simulation of the hard part is defined in the simulation, not the analysis and discussion of the soft part. The hard part is defined as ABS plastic, whose density is 1.06E–06 kg/mm^3, Young's modulus is 2.24 Gpa, Poisson's ratio is 0.38, yield strength is 20 Mpa, and ultimate tensile strength is 29.6 Mpa. Then the exoskeleton needs to be restrained and to simulate better what it would look like to be worn on a human body, the shoulders of the exoskeleton are completely fixed, with only horizontal degrees of freedom reserved for the ribs on both sides. The next step is to add the load, the force applied to the three points behind the simulated bandage, and the tightening force required for the chest guard (Fig. 2.5). The meshing of the model is based on the calculation of the software (Fig. 2.6). The average element size of the grid is divided based on 1–10% of the model size. After completing the above steps, it will be handled by the cloud. Three loads were placed behind the exoskeleton: the right shoulder, the left rib, and the right hip. In order to simulate the real situation of wearing force, the front chest protector used a fully fixed way to

Fig. 2.5 The exoskeleton model after the simplified model, material definition, constraint, and load addition

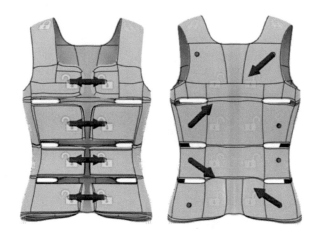

Fig. 2.6 Corrected exoskeleton after gridding

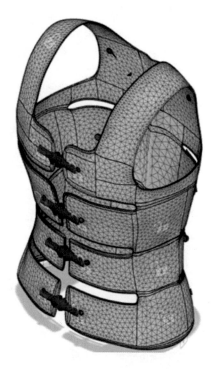

constrain the two sides, and the constraint position was evenly arranged. The three loads on the front side simulate the squeezing effect of the body on the protective gear. Here, a load of 10 N is applied to each extrusion point.

According to the final results, the safety factor is 15; the maximum Mises equivalent stress is 2.924 Mpa, and the minimum Mises equivalent stress is 0 Mpa (Fig. 2.7); the maximum displacement is 0.02453 mm, and the minimum displacement is 0 mm (Fig. 2.8); the maximum reaction force is 1.552 N, and the minimum reaction force is 0 N (Fig. 2.9).

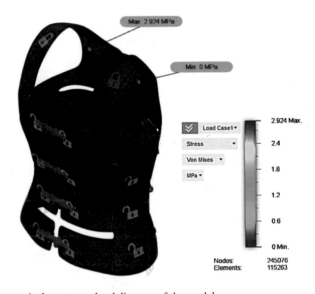

Fig. 2.7 Mises equivalent stress cloud diagram of the model

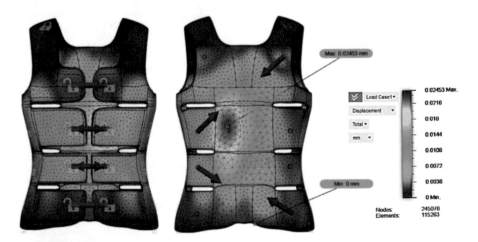

Fig. 2.8 Deformation cloud diagram after stress

Fig. 2.9 Cloud diagram of reaction force after stress

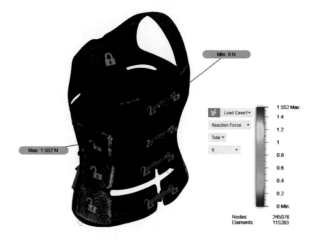

The method of finite element analysis is to use mathematical methods to simulate the real geometry and load conditions, and to approach the actual system with infinite unknowns with a limited number of tiny elements. Its unique characteristics and advantages make it irreplaceable in the study of scoliosis biomechanics. The application of finite element analysis in the treatment of scoliosis dates back to 1986, Viviani et al. [10] used finite element analysis to simulate scoliosis surgery. It can solve the practical problems that patients encounter in data collection, can save a lot of expensive economy on the animal experiment cost and time cost, and the mechanical properties of the brace structure are significantly improved after finite element analysis, because of the structure was optimized, the weight of the brace is to reduce it to a certain extent, enhance the comfort of patients.

2.5 Conclusion

In this paper, a design method of rehabilitation exoskeleton for scoliosis patients and its static analysis simulation test are proposed. According to the static simulation results of the finite element, the axial and horizontal correction forces used by the brace can obtain the continuous and effective correction force to eliminate the curve, and can effectively control the aggravation of the deformity and correct scoliosis. And a large number of clinical cases have proved that brace therapy is the only effective conservative therapy for idiopathic scoliosis.

Funding The study was supported by Key projects of national key research and development plan (2017YFF0207400): Research on key technologies and important standards of health services and remote health monitoring for the elderly and the disabled.

References

1. Taylor, H. J., Harding, I., Hutchinson, J., et al. (2013). Identifying scoliosis in population-based cohorts: development and validation of a novel method based on total-body dual-energy x-ray absorptiometric scans. *Calcified Tissue International, 92*(6), 539–547.
2. Khoshhal, Y., Jalali, M., Babaee, T., & et al. (2019). The effect of bracing on spinopelvic rotation and psychosocial parameters in adolescents with idiopathic scoliosis. *Asian Spine Journal, 13*(6). https://doi.org/10.31616/asj.2018.0307.
3. Grant, A. (2016). Letter to the Editor: The effectiveness of the SpineCor Brace for the conservative treatment of adolescent idiopathic scoliosis. Comparison with the Boston Brace. *The Spine Journal, 16*(8):1028–1029.
4. Zaina, F., De Mauroy, J. C., Grivas, T., et al. (2014). Bracing for scoliosis in 2014: state of the art. *European Journal of Physical and Rehabilitation Medicine, 50*(1), 93–110.
5. Mauroy, J. C. D., Pourret, S., Frédéric, B. (2016). Immediate in-brace correction with the new Lyon brace (ARTbrace): Results of 141 consecutive patients in accordance with SRS criteria for bracing studies. *Annals of Physical and Rehabilitation Medicine, 59*, e32.
6. Babaee, T., Kamyab, M., Ahmadi, A., Sanjari, M. A., Ganjavian, M. S. (2019). The intra- and inter-observer reliability of interface pressure measurements of the Milwaukee brace in adolescents with hyperkyphosis. *Journal of back and musculoskeletal rehabilitation, 32*(4).
7. Tu, Z., Wang, B., Li, L., et al. (2018). Early experience of full-endoscopic interlaminar discectomy for adolescent lumbar disc herniation with sciatic scoliosis. *Pain Physician, 21*(1), E63–E70.
8. Raison, M., Ballaz, L., Detrembleur, C., et al. (2012). Lombo-sacral joint efforts during gait: Comparison between healthy and scoliotic subjects. *Studies in Health Technology and Informatics, 176*, 113–116.
9. Cobetto, N., & Carl-Éric Aubin, P. S. et al. (2017). 3D correction of AIS in braces designed using CAD/CAM and FEM: a randomized controlled trial. *Scoliosis and Spinal Disorders, 12*(1), 24.
10. Viviani, G. R., Ghista, D. N., Lozada, P. J., et al. (1986). Biomechanical analysis and simulation of scoliosis surgical correction. *Clinical Orthopaedics and Related Research, 208*, 40–47.

Chapter 3
Simulation and Research of Upper Limb Rehabilitation Evaluation System Based on Micro Inertial Sensor Network

3.1 Introduction

With the advent of aging society in various countries, more and more geriatric diseases have increased year by year [1, 2]. Stroke is one of leading diseases causing adult death and disability worldwide [3]. According to the principles of neurodevelopment and neurophysiology, neuro-plasticity and functional reorganization [4], stroke patients can gain active control of limbs and improve the ability of limb activity if they can carry out limb rehabilitation training in time. Rehabilitation evaluation is one of the most important links in the whole rehabilitation treatment. At present, there are many methods to evaluate the patients' motor function in the world, such as Brunnstrom [5], Fugl-Meyer [6] and so on. These evaluation methods are all artificial evaluation, and the results are obtained by subjective judgment of rehabilitation doctors. Although this method is widely used in clinical practice, there are many uncertain factors that cause the following problems [7]: Firstly, most rehabilitation doctors cannot participate in the whole training process in real time, so that the evaluation results are mostly based on the experience of doctors and short-term observation, and the accuracy of rehabilitation evaluation depends on the experience level of doctors; Secondly, it is difficult for doctors to evaluate the long-term rehabilitation effect of patients at home and in the community, which will result in the problem of over training; Thirdly, most of the current commonly used evaluation scales do not change the scoring standards during the entire rehabilitation process, but with the increase of rehabilitation time, it is not the same that patients' basic ability of physical activity at the beginning of each rehabilitation period, and there is no multi-level scoring standard evaluation quantity. The results cannot accurately reflect the rehabilitation process and effect.

In order to solve this problem, many teams have proposed their own solutions. Most teams have designed a specific evaluation system on a specific rehabilitation exoskeleton or other specific equipment. Although they solved the problem of inaccurate evaluation when using rehabilitation devices, it is narrow that the scope of application and cannot help the rehabilitation training process in more scenarios. In order to solve the problem of evaluation accuracy in more situations, we designed an upper extremity rehabilitation evaluation system based on micro inertial sensor network, and used a variety of quantitative analysis methods to comprehensively evaluate rehabilitation training [8].

3.2 Methods

3.2.1 System Overview

In this paper, the rehabilitation upper limb exoskeleton system is used. As shown in Fig. 3.1, it is an exoskeleton instrument used for patient upper limb rehabilitation training. The Fig. 3.2 shows the micro inertial sensors. In addition, the system has data processing software and graphical user interface developed by Matlab. This paper is to simulate and study the evaluation of the system, from the perspective of motion, motion visualization [9], motion matching [10] three aspects of evaluation, and finally integrate it to achieve a comprehensive evaluation. These three aspects mainly relay on three position data (in the world coordinate system) and four quaternion data (in the world coordinate system), that is, the position, the rotation and other attitude information of each part of the upper limb. These data are acquired by the inertial sensor network through a specific motion capture algorithm [11, 12].

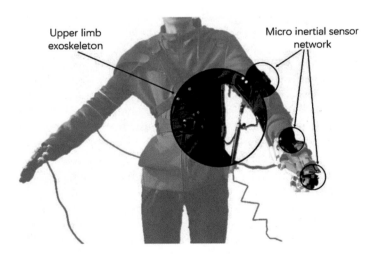

Fig. 3.1 Rehabilitation exoskeleton with upper limb rehabilitation evaluation system

Fig. 3.2 Micro inertial sensor

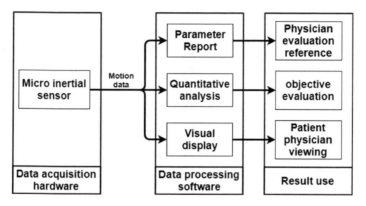

Fig. 3.3 Rehabilitation assessment system flow

Through a certain algorithm to calculate the motion angle, through the DTW algorithm to compare and evaluate the patient action and the standard action, through MATLAB to achieve the visual simplification of the action. As shown in Fig. 3.3, the system program includes data retrieval and processing from micro inertial sensor network, simplified human body model, motion angle calculation, activity matching score, contrast score and visual display. The following is to describe each link in detail.

3.2.2 Simplified Model Building of Upper Limb Multi-body System

There are 206 bones in the human body that we can see from the human bone structure [13], and they are driven by the ligaments connecting the muscles. The bone is used as the evaluation object of rehabilitation training. If every bone is calculated, the whole system model will be complicated. Therefore, we simplify 206 human bones

Fig. 3.4 Figure of human line

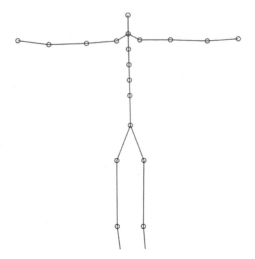

into 20 through the analysis of bone movement. As shown in Fig. 3.4, the limb part (4 parts in total) sets each joint as a node, with a simplified bone between nodes and a simplified bone at the end. From the hip to the head, the nodes are set at a certain interval between the lumbar and cervical vertebrae, and there is a simplified bone between each node. The connection of the five parts is shown in Fig. 3.4. Each simplified skeleton is a rigid body, and the motion between adjacent rigid bodies is mutually affected.

3.2.3 Analysis and Processing Method of Upper Limb Posture Information

We extract the acceleration and angular velocity information from the micro inertial sensor, and calculate the displacement information by quadratic integration of the acceleration information, and use this information to express the position information of 20 bones.

$$P_T = P_0 + \int_0^T v_0 dt + \iint_0^T a_t dt^2 \qquad (3.1)$$

where P is displacement, V is velocity, a is acceleration, T is termination time, 0 is initial time, and t is intermediate time.

Then the angular velocity information is transformed into Euler angle information, and then transformed into quaternion information through Euler angle. The Euler angle is obtained by using the angular velocity. The Euler angle is obtained in the order of Z, Y and X. the rotation angles of the object around X, Y and Z are respectively recorded as φ, θ and ω. Define quaternion $q = [W\ X\ Y\ Z]'$, among $|q^2| = W^2 + X^2 + Y^2 + Z^2 = 1$. Use the formula 4 to get quaternion information. Then:

$$q = \begin{bmatrix} W \\ X \\ Y \\ Z \end{bmatrix} = \begin{bmatrix} \cos(\varphi/2)\ \cos(\theta/2)\ \cos(\psi/2) + \sin(\varphi/2)\ \sin(\theta/2)\ \sin(\psi/2) \\ \sin(\varphi/2)\ \cos(\theta/2)\ \cos(\psi/2) - \cos(\varphi/2)\ \sin(\theta/2)\ \sin(\psi/2) \\ \cos(\varphi/2)\ \sin(\theta/2)\ \cos(\psi/2) + \sin(\varphi/2)\ \cos(\theta/2)\ \sin(\psi/2) \\ \cos(\varphi/2)\ \cos(\theta/2)\ \sin(\psi/2) - \sin(\varphi/2)\ \sin(\theta/2)\ \cos(\psi/2) \end{bmatrix}$$

$$(3.2)$$

Based on the position information and the quaternion rotation information, we used MATLAB to express the simplified human figure, that is, to reproduce the motion of the patient training process.

3.2.4 Calculation of Motion Angle

(1) Mathematical modeling

In the muscle movements of the human upper limbs, there are six joints with kinematic pair effect, left shoulder joints, right shoulder joints, left elbow joints, right elbow joints, left wrist joints, left wrist joints, (six joints in total), and each joint has corresponding actions [14] (see Table 3.1). In addition to the joint as a kinematic pair, the internal and external rotation of the forearm is caused by the offset of the ulna and radius, which is driven by the corresponding muscle group. Each action in the human body has a specific range of activity, and the activity ability of the limbs is also determined by the range of activity. In order to facilitate the construction and measurement of the model and meet the needs of rehabilitation doctors, this paper studies this, and adopts the world coordinate system and its internal joint coordinate system to express the position, kinematic parameter, and range of activity of each joint.

According to the requirements of Brunnstrom, Fugl-Meyer (FMA) and other evaluation methods commonly used in the world. The evaluation of upper limb activity ability is to evaluate the activity ability of each joint, that is, the movement angle and position change of each joint in the process of movement. Therefore, the mathematical model of the whole upper limb is built as shown in Fig. 3.5, in which six local coordinate systems are used, including the left/right shoulder, left/right elbow and left/right wrist, and a world coordinate system. The position of the coordinate system is related to the attitude data obtained by the micro inertial sensor network, and the corresponding position is shown in Fig. 3.6. The origin of the world coordinate system is determined by the initial parameter value of the micro inertial sensor network, which mainly provides a unified intermediate coordinate system for each local coordinate to express the attitude. As shown in Fig. 3.5, the red coordinate

Table 3.1 Corresponding table of upper limb joints and movements

Joint	Action
Shoulder joint	Shoulder flexion
	Shoulder extension
	Shoulder horizontal abduction
	Shoulder horizontal adduction
	Shoulder abduction
	Shoulder adduction
	Shoulder medial rotation
	Shoulder lateral rotation
Elbow joint	Elbow flexion
	Elbow extension
(Elbow joint involved)	Forearm pronation
	Forearm supination
Wrist joint	Wrist extension
	Wrist flexion
	Wrist abduction
	Wrist adduction

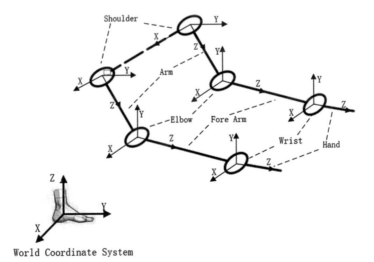

World Coordinate System

Fig. 3.5 Upper limb mathematical model building

system is the world coordinate system, with the right foot as the reference. Taking the shoulder coordinate system as the first coordinate system, they are arranged in order along the shoulder-arm-fore arm-hand, and the Z-axis direction is upward along the arm-anti-wrist direction.

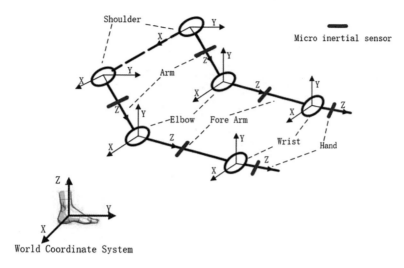

Fig. 3.6 Micro inertial sensor and coordinate system

The movement of shoulder joint can be simplified into three degrees of freedom: flexion and extension, horizontal adduction and abduction, medial rotation, and medial adduction and abduction. In order to express the action function of shoulder more clearly and simply, three orthogonal axes are chosen as the basic axis. Figure 3.7

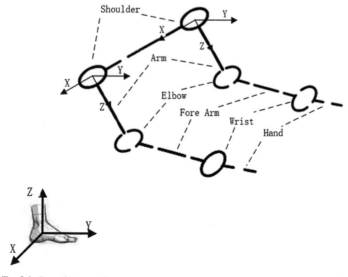

Fig. 3.7 Shoulder coordinate system representation

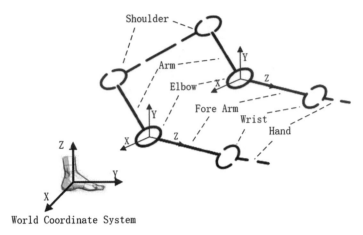

Fig. 3.8 Representation of elbow coordinate system

shows a simplified body line diagram of a system with coordinates. The corresponding X-axis is the roll axis, Y-axis is the heading axis, Z-axis is the pitch axis, the left shoulder coordinate system is named L_SH, and the right shoulder coordinate system is named R_SH.

The elbow joint has the least movement, but the highest proportion in the evaluation of the whole upper limb activity ability, less degree of freedom. In this model, the flexion and extension are simplified into one degree of freedom, and the forearm pronation involving elbow joint is simplified to one degree of freedom of elbow joint. There are two degrees of freedom in the elbow joint, and three orthogonal axes are selected as the basic axis. As shown in Fig. 3.8, the two degrees of freedom are respectively around the Z-axis and the Y-axis, and the left and right elbow coordinate systems are named L_EB and R_EB respectively.

The largest movement of wrist joint is flexion and extension, which can be simplified to a degree of freedom. Although the range of wrist horizontal adduction and abduction is small, it is still an important reference action to judge the upper limb movement ability, which is simplified to a degree of freedom. The wrist joint has two degrees of freedom, three orthogonal axes are selected as the basic axis. Figure 3.9 shows the position of the local coordinate system of the wrist joint in the simplified body line diagram. The left and right wrist coordinate systems are named L_WS and R_WS respectively.

(2) Calculation of motion angle

The mathematical model at the joint is shown in Fig. 3.10, where ST1 and ST2 represent two bones connected with joint J, and UJ and NJ are joints connected with ST1 and ST2 bones respectively. Among them, the quaternion is generated by the micro sensor and built in the world coordinate system, that is, the world coordinate system as shown in Fig. 3.10. At the same time, each joint also has its corresponding position local coordinate system, in which there are corresponding joint coordinate

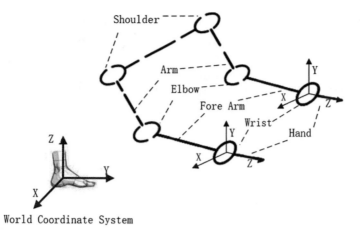

Fig. 3.9 Representation of wrist joint coordinate system

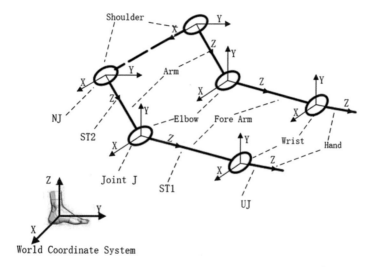

Fig. 3.10 Mathematical model for calculation of motion angle

points under the world coordinate system. Two sets of motion angles can be generated from two sets of data, namely θ and α.

Method of generating angle from position coordinate:

Set the length of ST1 as L1, ST2 as L2, and the coordinate vectors of joint J, Q1 and Q2 in the world coordinate system are respectively $Q_1 = (a_{1j})(j = 1, 2, 3)$, $Q_2 = (b_{1j})(j = 1, 2, 3)$, $J = (c_{1j})(j = 1, 2, 3)$.

That

$$\alpha = \arccos\left(\frac{L_1^2 + L_2^2 - \sqrt{\sum_{j=1}^{3}(b_{1j} - a_{1j})^2}}{2L_1 L_2}\right) \tag{3.3}$$

Generation angle method of quaternion:

ST2 produces two quaternions $v = [0, \mathbf{v}]$ and $v' = [0, \mathbf{v'}]$ before and after a rotation. The vector of ST2 in the world coordinate system is $\mathbf{u}$, set $q = \left[\cos\left(\frac{1}{2}\theta\right), \sin\left(\frac{1}{2}\theta\right)\mathbf{u}\right]$. According to $v' = qvq^* = qvq^{-1}$, θ can be obtained.

3.2.5 Evaluation and Visual Comparison of Upper Limb Activity Capacity

At present, Brunnstrom, Fugl-Meyer and other scales are mostly used for rehabilitation training evaluation of stroke patients in the world. As shown in Table 3.2, it is Brunnstrom semi quantitative table. The unified feature of these scales is that

Table 3.2 Brunnstrom semi quantitative table

Stage	Leg	Arm	Hand
1	Flaccidity	Flaccidity	No hand function
2	Spasticity develops, minimal voluntary movements	Beginning development of spasticity limb synergies or some of their components begin to appear as associated reactions	Gross grasp beginning, minimal finger flexion possible
3	Spasticity peaks, flexion and extension synergy present, hip-knee-ankle flexion in sitting and standing	Spasticity increasing; synergy patterns or some of their components can be performed voluntarily	Gross grasp, hook grasp possible, no release
4	Knee flexion past 90° in sitting, with the foot sliding backward on the floor, dorsiflexion with the heel on the floor and the knee flexed to 90°	Spasticity declining; movement combinations deviating from synergies are now possible	Gross grasp present, lateral prehension developing, small amount of finger extension and some thumb movement possible
5	Knee flexion with the hip extended in standing, ankle dorsiflexion with the hip and knee extended	Synergies no longer dominant, more movement combinations deviating from synergies performed with greater ease	Palmar prehension, spherical and cylindrical grasp and release possible
6	Hip abduction in sitting or standing, reciprocal internal and external rotation of the hip combined with inversion and eversion of the ankle in sitting	Spasticity absent except when performing rapid movement, isolated joint movements performed with ease	All types of prehension, individual finger motion, full range of voluntary extension possible

Table 3.3 Brunnstrom full quantitative table

Stage	Name	Condition	Parameter
I	Forearm	Round surface	Angle range <5°
II	Forearm		Angle range 5° < A
III	Elbow angle		Angle range ≤10°
IV	Forearm angle Hands on the back Elbow angle Rotate the forearm		≤10° 180°–90° ≥90°
V	Forearm angle Shoulder angle Shoulder and vertical angle Elbow angle 180°	Forearm rotation >90°	≤5° 90° $\arctan\left(\frac{Head\,lengh}{Arm\,lengh}\right)$
VI	Forearm wobble angle		≤2°

they are all semi quantitative tables, which can't be judged by computer program, and can only be judged by subjective experience of rehabilitation doctors. Therefore, we converted the upper limb part of the Brunnstrom semi-quantitative scale into a full-quantity scale together with an experienced rehabilitation physician, as shown in Table 3.3, which is the full quantitative table of Brunnstrom upper extremity part.

Through the calculation method of motion angle, it is obtained that the parameters needed in the quantitative table. The ratio of the obtained evaluation parameters to the standard evaluation parameters in the quantitative table is calculated. Weighted according to the comparison value of the latest evaluation results. The final evaluation score is obtained after calculation according to the formula.

Assume that the assessment angle of elbow flexion angle of a level IV patient is α, and the last evaluation score is e (out of 10). According to Brunnstrom full quantitative table and scoring method, the scores can be:

$$E = \left(\tfrac{\alpha}{90} + 1\right)e \qquad (3.4)$$

At the same time, the data curve of the patient's movements and the standard data curve provided by the rehabilitation physician are displayed in the same graph, so that the rehabilitation doctors can refer to the whole movement process.

3.3 Result

Base on theoretical knowledge, we discussed the actual effect data of a stroke patient in the stage IV of rehabilitation training. In the stage IV of rehabilitation training, the stroke patient will gradually appear the initiative to break the common action, the spasm will weaken and tend to disappear, that is, the muscle tension will decrease. At this time, the rehabilitation doctors mainly observe and evaluate the three movements

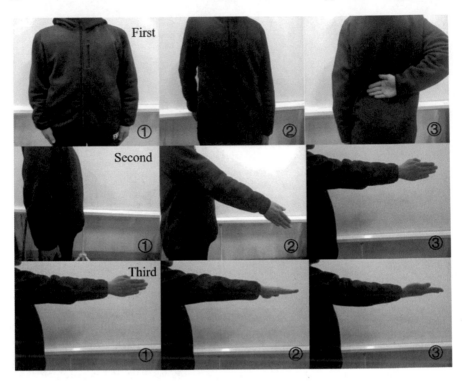

Fig. 3.11 flexion and extension process of single elbow

of the patients, namely: whether the hand can be extended backward and touch the waist; whether the hand can be raised forward to ninety degrees; forearm pronation and forearm supination.

We recorded the movement process of the patient through the micro inertial sensor network, as shown in Fig. 3.11. This is the three main fixed movements in the upper limb rehabilitation evaluation, which are mainly displayed as the movement angle, movement displacement range, and rotation angle.

As shown in Fig. 3.12, the visual motion graph is divided into four areas. The upper left corner is a three-dimensional view of the body line chart, which can reproduce the whole training process of the patient; the upper right corner shows the main view mainly from the front to view the patient's actions; the lower left corner shows the side view mainly from the side to display the patient's actions; the lower right corner shows the top view. In the choice of perspective, we analyzed and evaluated the different observation angles of the human body. At present, we mainly focused on the evaluation of the upper limb rehabilitation training results. The axonometric map shows the complete actions of the patients in the rehabilitation training process, and the three views respectively show the angular range, vertical displacement range and horizontal displacement range of the upper limb movement. The purpose is to

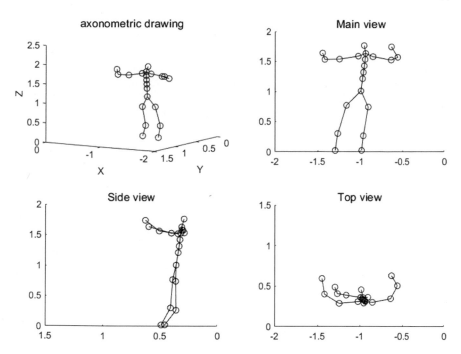

Fig. 3.12 Multi perspective representation of human visual motion graph

record and simply show the patient's training process, it is convenient for doctors to observe the whole process of patient training and evaluate more accurately.

As shown in Figs. 3.13 and 3.14, the action curve comparison chart shows the difference between standard action and patient action. In Fig. 3.13, the ordinate represents the displacement difference between the patient's upper limb movement and the standard movement during the process of motion, and the abscissa represents the number of frames. In Fig. 3.14, the Ordinate represents the Euler angle of the patient's upper limb movement during the process of motion, and the abscissa represents the number of frames.

Because the movement angle and displacement of the upper limb are certain, the rehabilitation effect can be evaluated by analyzing and comparing the data of patients' fixed movement. As the normal upper limb movement such as displacement and rotation angle have a certain range, the rehabilitation training score is calculated by comparing the patient's movement data with the normal upper limb movement data, and weighted evaluation with the rehabilitation evaluation suggestions provided by the rehabilitation doctor.

From the above two pictures, the recovery of this stage IV stroke patient is good. The displacement difference in Fig. 3.13 decreases with the movement process, and gradually approaches the movement range of the normal upper limb. In Fig. 3.14, the deviation degree of the two movement angle curves can also verify the rationality of

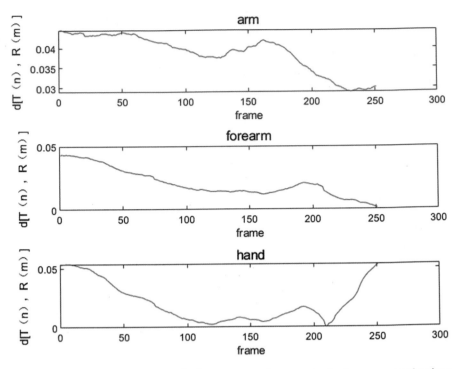

Fig. 3.13 Comparison of movement displacement curve between standard movement and patients

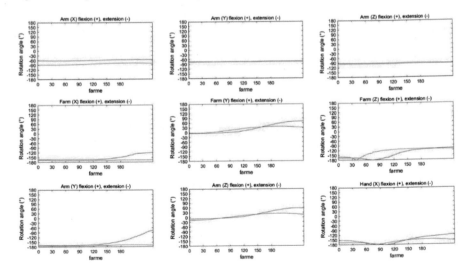

Fig. 3.14 Comparison of angle curve between standard action and patient

the patient's rehabilitation. After communication with the rehabilitation doctor, the conclusion is also consistent.

3.4 Conclusions and Future Work

In order to reduce the instability brought by the subjective evaluation of rehabilitation doctors, it is necessary to measure and calculate the limb movement data of patients and increase the proportion of accurate data in the evaluation. In normal movement, such as the upper limb, the movement angle between the big arm and the small arm is certain. By comparing the calculation of the movement angle with the range of the normal movement angle, we get the disability level and recovery degree of the patient's limbs. After exploring the actual effect of a certain rehabilitation training data of a stage IV stroke patient, and analyzing and evaluating the generated movement curve comparison diagram and visual figure diagram, we get the limb movement angle of the patient, and compare and score the accurate and developed full quantitative table. Compared with the evaluation results of the rehabilitation physician, there is a slight lack of language description level, but the data level is more accurate than the description of rehabilitation doctors, which basically meets the needs of rehabilitation evaluation. It solves the problems of imperfection and inaccuracy caused by the single use of Rehabilitation Exoskeleton Robot and the subjective evaluation of rehabilitation doctors, and provides suggestions for the rehabilitation training later, avoiding the possible secondary injury caused by over training.

After the basic construction of the system, carried out a certain amount of simulation and experiment. Through the existing experimental results, we can clearly get the movement data of the patient's affected limb. According to the analysis and processing of the obtained movement data, we get the data required by the rehabilitation doctors in the subjective evaluation, and further improve the efficiency of rehabilitation training. At present, the experiment only aims at the collection and analysis of upper limb motion data, but it has reliable experimental results and theoretical support in further expanding the function of whole limb motion evaluation. Through the data collection of displacement, acceleration and Euler angle generated in the fixed rehabilitation assessment, and then compared with the range of motion data of standard motion, the recovery degree of patients can be simply obtained by generating the data comparison curve. So only need to improve the assessment method of the whole limb to achieve the goal. Due to the independence and connectivity of the system, it is suitable for different Rehabilitation Exoskeleton robots. The main data acquisition method of the system is to use micro inertial sensor network, and the independent wearing method can be adapted to most of the Rehabilitation Exoskeleton robots. It provides a guarantee for its subsequent commercial value and feasibility.

Funding The study was supported by Key projects of national key research and development plan (2017YFF0207400): Research on key technologies and important standards of health services and remote health monitoring for the elderly and the disabled.

References

1. Yi, M. (2007). Research report on prediction of population aging development trend in China. *China Women's Movement, 2,* 17–20.
2. Wu, Q. (2017). State Council: release the 13th five-year plan for the development of national aging cause and the construction of pension system. *China Business, 4,* 74.
3. Chen, J., Liu, M., Sun, D., et al. (2018). Effectiveness and neural mechanisms of home-based telerehabilitation in patients with stroke based on fMRI and DTI: A study protocol for a randomized controlled trial. *Medicine, 97*(3), e9605.
4. Piradov, M. A., Chernikova, L. A., & Suponeva, N. A. (2018). Brain plasticity and modern neurorehabilitation technologies. *Herald of the Russian Academy of Sciences, 88*(2), 111–118.
5. Liparulo, L., Zhang, Z., Panella, M., Gu, X., & Fang, Q. (2017) A novel fuzzy approach for automatic Brunnstrom stage classification using surface electromyography. *Medical & Biological Engineering & Computing, 55*(8), 1367–1378.
6. Rech, K. D., Salazar, A. P., Marchese, R. R., Schifino, G., Cimolin, V., Pagnussat, A. S. (2020). Fugl-Meyer assessment scores are related with kinematic measures in people with chronic hemiparesis after stroke. *Journal of Stroke and Cerebrovascular Diseases: The Official Journal of National Stroke Association, 29*(1), 104463.
7. Tongyang, S., Hua, L., Quanquan, L., et al. (2017). Inertial sensor-based motion analysis of lower limbs for rehabilitation treatments. *Journal of Healthcare Engineering, 2017,* 1–11.
8. Wang, A. M. (2009). *Evaluation of rehabilitation function.* Shanghai: Fudan University Press.
9. Oshita, M., Inao, T., Ineno, S., et al. (2019). Development and evaluation of a self-training system for tennis shots with motion feature assessment and visualization. *The Visual Computer, 35*(3), 1–13.
10. Misu, S., Asai, T., Doi, T., et al. (2019). Development and validation of Comprehensive Gait Assessment using InerTial Sensor score (C-GAITS score) derived from acceleration and angular velocity data at heel and lower trunk among community-dwelling older adults. *Journal of NeuroEngineering and Rehabilitation, 16*(1), 62. https://doi.org/10.1186/s12984-019-0539-3.
11. Dao-Xiong, G., Rui, H. E., Guo-Yu, Z., et al. (2018). Motion mapping in the joint space for the control of the heterogeneous wheelchair-mounted robotic arm. *Acta Electronica Sinica, 46*(2), 464–472.
12. Lisini, T., Farina, F., Garulli, A., et al. (2020). Upper body pose estimation using wearable inertial sensors and multiplicative Kalman filter. *IEEE Sensors Journal, 20*(1), 492–500.
13. Weingaertner, T., Hassfeld, S., Dillmann, R. (1997). Human motion analysis: A review. In *IEEE Workshop on Motion of Non-rigid & Articulated Objects* (Vol. 1, pp. 0090). IEEE Computer Society.
14. Crompton, R. H., Wood, B., Cummings, S. W., et al. (2018). Encyclopædia Britannica. Human muscle system [DB/OL]. Encyclopædia Britannica, inc. https://www.britannica.com/science/human-muscle-system.

Chapter 4
Design and Simulation of Controllable Soft Drive for Exoskeleton Robot

4.1 Introduction

Adolescent idiopathic scoliosis is a common condition found in adolescents. For this kind of spinal deformity, it is usually necessary to use rigid brace as a treatment method. The discomfort caused by the rigid brace and the psychological distress caused by its appearance will have a negative impact on users [1]. The definition of adolescent idiopathic scoliosis by SRS is as follows: No definite cause, scoliosis occurs in patients with more than 10° scoliosis from the age of 10 to skeletal maturity whether or not accompanied by sagittal or cross-sectional changes. Bracing can be defined as the application of external corrective forces to the trunk; rigid supports or elastic bands can be used, and supports can be customized or prefabricated [2]. A TLSO is a type of brace used to control the lateral curvature of the spine. It is a nonsurgical treatment with the goal of preventing curve progression in patients with idiopathic scoliosis. It can generally evaluate and improve the compliance of adolescent idiopathic scoliosis patients [3]. However, it is still based on the common brace, which interferes with daily activities in use and is difficult to adapt to the changes in the spine. Computer-aided design (CAD) and computer-aided manufacturing (CAM) can be used to standardize specific styles of support processing and improve wearing comfort [4], but it can still bind most of the body. In this paper, a new scheme is proposed to incorporate the soft drive into the brace. It brings flexibility and dynamics to the stent, which can increase the body's mobility and reduce constraints on other parts of the body. The soft drive is designed primarily for PUMC concordance type I a scoliosis patient whose vertices are located in T2–T11, T12 lumbar spine.

4.2 Methodology

4.2.1 Soft Structure Design

The main part of the controllable soft drive that assists spinal correction is its soft part. It has three functions. One is to stretch the spine, the other is to reduce chest compression, and the third is to achieve adjustable traction.

The soft drive is made of food grade silica gel with high pressure and tensile resistance. Soft drive is inherently adaptability and lightweight [5]. It is resilient enough to bend without making the soft part of the drive appear too soft. In the process of working, it will not cause severe pressure to patients due to rigid impact. The principle of the actuator is as follows: under the same pressure, the elongation of materials of different thickness is different. As the pressure of the soft body cavity keeps increasing, the expansion rate of each chamber in the thinner part is much higher than that in the thicker part. Thus, the entire thicker portion is bent. The shape of the soft body is cylindrical, and each chamber can be deformed to correct scoliosis.

The spine has many degrees of freedom. The most common used four directions are selected to design the soft structure to complete the backward, left, right pulling and auxiliary bending process. It can also produce different sizes of correction forces on the spine in different directions. The four chambers realize the deformation in different directions. The four chambers are different in size. The largest one is on the side far away from the spine, and the two symmetrical chambers on the left and right bear the main pulling force. Under different air pressure, the bending angle of the soft drive is different. Therefore, the required traction can be controlled by the inflation time (Figs. 4.1 and 4.2).

Fig. 4.1 Soft structure diagram

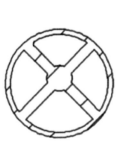

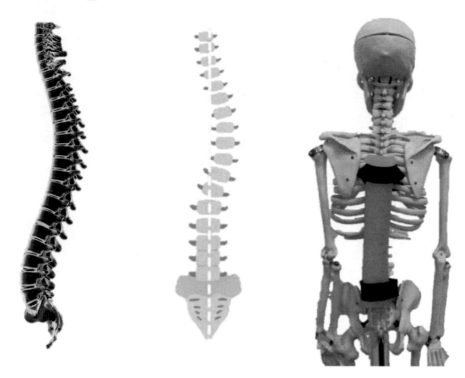

Fig. 4.2 Soft drive

4.2.2 Process and Control Framework

The whole control system is composed of three parts: sensor, control module and soft drive. The control part is mainly composed of microcontroller, compound transistor array, bending sensor, solenoid valve, etc. The control mode includes automatic and manual mode. The complete control system can achieve dynamic adjustment of force, in which the independent bending sensor and the soft drive are fixed near the disc T2 and the intervertebral disc T11–T12 respectively. When the patient's lumbar spine begins to bend, the bending sensor follows the lumbar spine bend; the cobb angle is passed to the control module; the control module converts the angle information into inflation time, and then control the internal pressure of the soft drive; the soft drive bends. The direction and the bending sensor transmit the angle signal in the opposite direction, and the soft drive pulls the spine. The bending sensor detects a change in the angle information and stops the inflating when the Cobb angle is 0°, thus correcting the lumbar vertebral bend (Fig. 4.3).

The control system adopts the manual mode by mobile remote control, which can perform auxiliary functions and self-adjust the pulling force. When the upper pneumatic cavity of the main body is inflated, the posterior cavity slowly expands toward the spine so as to achieve the effect of pulling backwards. When the left and

Fig. 4.3 Control system

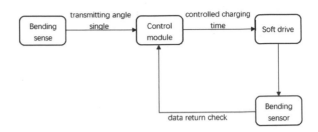

right pneumatic cavity are inflated separately, in a certain limit, the pneumatic soft body will expand to the left or right respectively, to achieve the pull effect correcting the bent spine. When the patient bends over, the lower air chamber will expand and the main body of the drive will protrude outward. Its shape is the same as that when bending the human body to help teenagers bend correctly.

4.3 Discussion of Results

4.3.1 Simulation

The finite element analysis in every process could be effort the theoretical guidance in process design and shorten the development cycle [6]. In order to test the effect and optimize the design of the soft drive, the finite element analysis of the soft drive was carried out by using the ABAQUS 6.14 Computer Aided Engineering (CAE) soft package (SIMULIA, Dassault systems) [7]. The controllable soft drive was the main executive component of the auxiliary correction of scoliosis. The bending angle and bending displacement of the controllable soft actuator would directly affect the effect of correction of scoliosis. For the calculation of stresses finite element method was used [8]. Therefore, in this paper, the finite element analysis of the actuator's bending displacement and stress-strain under different air pressures was carried out.

According to type I, scoliosis patients with scoliosis belonged to the large deformation of the spine, so the 3D model of the soft drive was suitable for Ogden when performing finite element analysis. On a qualitative aspect, the model deformed by Ogden's law followed the concavity of the reconstructed model [9]. According to the density of silica gel, the constitutive model was given corresponding material properties. There were six parameters to be determined in Ogden constitutive model. They were shear modulus mu, material coefficient alpha 2 and compressibility constant D. The specific experimental data of the three parameters to be collected were shown in the Table 4.1. Then the parameters were determined by data fitting.

In the original results, we need to find the stress-strain data to fit the mu and alpha parameters through the third-order Ogden constitutive model formula. The following Fig. 4.4 was the result of our data fitting.

Table 4.1 Measurement result sheet

Elongation	Area	Pull
50	8×10^{-5}	0
60	8×10^{-5}	6
70	8×10^{-5}	9
80	8×10^{-5}	13
90	8×10^{-5}	16
100	8×10^{-5}	19
110	8×10^{-5}	20
120	8×10^{-5}	26
130	8×10^{-5}	35

	mu1	alpha1	mu2	alpha2	mu3	alpha3	D1	D2	D3
1	1.211	1.0921	0.0004021	9.8098	5.611E-005	-10.055	0.4989	0.4989	0

Fig. 4.4 Parameter input diagram

And then we need to set the load. The pressure was set at 0.025 MPa and one Chambers were selected as the main Chambers for inflation. Then we need to set the constraints of this model to limit the movement of one of its faces in the x direction and the rotation in the y and z directions. And then we need to submit the analysis homework.

The simulation results show that when the pressure is 0.025 MPa, the stress-strain effect is better. The stress-strain relationship is almost proportional, and the bending Angle can reach 45°. The simulation results obtained from this point of view, from the perspective of rehabilitation medicine, soft drive can theoretically help the rehabilitation of patients with type I scoliosis (Figs. 4.5, 4.6, 4.7, Table 4.2).

Fig. 4.5 Side view, soft drive bending effect

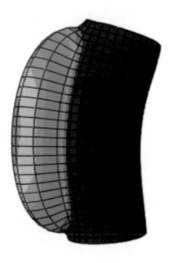

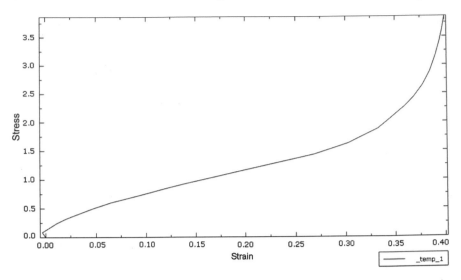

Fig. 4.6 Theoretical stress-strain diagram of elastic materials

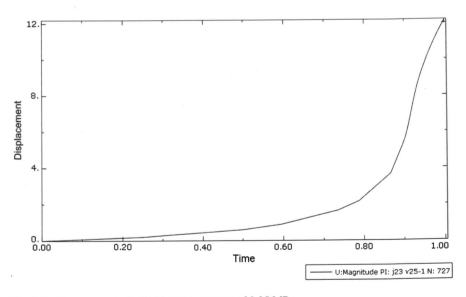

Fig. 4.7 Displacement of soft drive at a pressure of 0.25 MPa

4.3.2   Experiment

Firstly, we used a 3d printer to print the model. Secondly, we modulated the silica gel soft body at the appropriate proportion. Thirdly, we solidified the silicone in the

Table 4.2 Time and displacement

Time	0	0.25	0.5	0.59375
Displacement	0	0.207046	0.597959	0.881819
Time	0.734375	0.787109	0.866211	0.895874
Displacement	1.64986	2.15568	3.67941	5.23509
Time	0.90329	0.910706	0.921829	0.926001
Displacement	5.69845	6.38729	7.58933	8.02295
Time	0.932258	0.941643	0.955722	0.9698
Displacement	8.57747	9.25701	10.0989	10.8222
Time	0.983878	1		
Displacement	11.4629	12.1296		

mold. Finally, we sealed the soft drive and inserted it into the soft drive after post-processing (Fig. 4.8). An efficient and low-cost way was by using a microcontroller such as Arduino Nano [10]. During the charging process, the Arduino Nano board

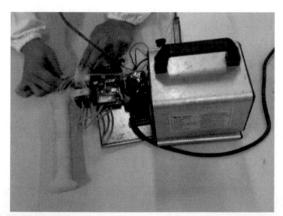

Fig. 4.8 Experimental equipment and gas path

controlled the switch of the air valve. Before entering the soft drive, the pressure-limiting valve limited the gas pressure which enter the soft drive. The program was burned to the Arduino board to control the switch of different air valves. Therefore, the inflation and venting of each chamber were controlled. In the case of inflation, the soft drive would have obvious expansion and bending. The pneumatic bending effect of silicone soft drive was good, which could achieve Ω bending degree of shape.

The self-designed controlled soft drive was used as the experimental material, and a suitable gas with different pressures was continuously injected into the soft drive to make it bend, and the bending was made arbitrarily at a certain Angle of 0–70° for data measurement.

We spread a piece of 630 mm × 297 mm square paper with the square size of 10 mm × 10 mm on the experiment table. The controlled soft drive was placed on the grid paper, one end was fixed. This fixed endpoint plane coordinate was recorded as point O (x1, y1) and the plane coordinate system was established on the grid paper with point O as the origin. Then the information of the plane coordinate point P (x2, y2) at the non-fixed end was recorded. When the pressure inside the driver was equal to the pressure of inflation, the driver reached the maximum bending Angle and maintains the shape unchanged. At this time, the information of the coordinate point Q (x3, y3) at the non-fixed end was recorded.

It could be easily obtained from the trigonometric functions:
Bending Angle is

$$\alpha = \arcsin\left(\frac{x3 - x1}{y3 - y1}\right) \tag{4.1}$$

It could be easily obtained from the two-point coordinate formula:
Bending displacement is

$$s = \sqrt{(x_3 - x_2)^2 + (y_3 - y_2)^2} \tag{4.2}$$

A set of experimental bending angle parameters and bending displacement parameters were obtained. Kept other conditions unchanged and repeated the experiment for several times. Then the coordinate plane information of corresponding points O, P and Q was recorded in the same way.

By calculating bending Angle parameters and bending displacement parameters in each group, it could be obtained that the driver could reach 0–90° of bending Angle and displacement of 0–152 mm length in the real experiments. From the angle of rehabilitation medicine, the controlled soft drive on the experiment could be proved to help type I a scoliosis patient with the rehabilitation of the spine (Fig. 4.9 and Table 4.3).

Therefore, when the air pressure was 0.025 MPa, the real displacement of the 175 mm long soft drive was 69 mm. First, we drew a straight line $\overline{OC}$, which represents the soft drive in the uninflated state, with a length of 175 mm. Second

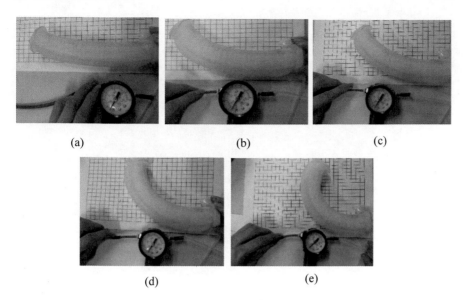

Fig. 4.9 Bending change of soft drive under different time

Table 4.3 Data sheet

Pressure (MPa)	0	0.01	0.02	0.025	0.03	0.04
Actual displacement (mm)	0	0	26	69	107	152

we drew a circle with a radius of 69 centering on point C. Third, we made arc $\widehat{OC_1} = 175$ mm and tangent to straight line $\overline{OC}$, where $\widehat{OC_1}$ represents the soft drive under 0.025 MPa. Fourth, we limit the center of arc $\widehat{OC_1}$ to a straight line perpendicular to $\overline{OC}$, 175 mm away from point C. It could be measured that $\angle 1 = 46.0°$, and finally we drew a straight line $\overline{OA} = 75$ mm, where $\overline{OA}$ was the length of soft drive under simulation. As the ratio of arc $\widehat{OA_1}$ to $\widehat{OC_1}$ and the ratio of $\angle 1$ to $\angle 2$ were equal, from $\frac{\angle 1}{\angle 2} = \frac{\widehat{OC_1}}{\widehat{OA_1}}$, it could be obtained that $\angle 2$ was 19.7°, and the line $\overline{AA_1} = 12.2080$ mm. In 3.1, the simulation analysis was carried out for the soft drive with a length of 75 mm. From Table 4.2, it could be seen that the displacement of the soft drive was 12.1279 mm when the air pressure was 0.025 MPa. It could be seen that the simulation results were basically consistent with the actual displacement, and the error rate was $\frac{|12.1296-12.2080|}{12.2080} = 0.64\%$. Therefore, it was effective to use the finite element analysis method for the soft drive, and the simulation results were accurate. According to Fig. 4.10, $\angle 1 = 46.0°$, which was the Cobb angle with the largest bending angle of the soft drive under the air pressure of 0.025 MPa. Therefore, under the pressure of 0.025 MPa, the soft drive can help the type I a scoliosis patient whose Cobb angle reaches 46° to have correction and rehabilitation.

Fig. 4.10 Computation
figure

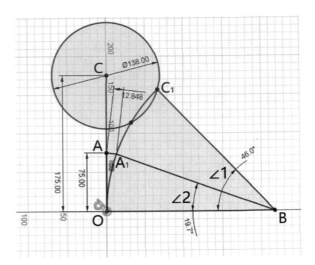

4.4 Conclusion

This paper proposes a feasible scheme for scoliosis correction and the key part of the soft drive is simulated. Abaqus' simulation experiments on the soft drive verify the rationality of its internal structure. The experimental results show that the designed soft drive can achieve a bending angle of 46.0°, which can achieve a certain degree of scoliosis correction. Finally, a feasible method for correction of scoliosis is proposed. The possibility of the soft drive to correct the spine is verified by experiments. By inflating each chamber of the soft to observe the bending situation, the expansion in the transverse direction may have an impact on the elongation and bending. In the future, new finite element models and soft-drive structures will be built to find that structures can withstand greater forces and deliver greater forces to help to correct the spine.

Funding The study was supported by Key projects of national key research and development plan (2017YFF0207400): Research on key technologies and important standards of health services and remote health monitoring for the elderly and the disabled.

References

1. Law, D., Cheung, M. C., Yip, J., et al. (2016). Scoliosis brace design: influence of visual aesthetics on user acceptance and compliance. *Ergonomics, 60*(6), 1–28.
2. Negrini, S., & Grivas, T. B. (2010). Introduction to the "Scoliosis" journal brace technology thematic series: increasing existing knowledge and promoting future developments. *Scoliosis, 5*(1), 2.

3. Bache, B. A., Iftikhar, O., Dehzangi, O. (2017). Brace treatment monitoring solution for idiopathic scoliosis patients. In *2017 16th IEEE international conference on machine learning and applications (ICMLA)*, Cancun, pp. 580–585. https://doi.org/10.1109/icmla.2017.00-98.
4. Weiss, H. R., Tournavitis, N., Nan, X., et al. (2017). Workflow of CAD/CAM scoliosis brace adjustment in preparation using 3D printing. *The Open Medical Informatics Journal, 11*(1), 44–51.
5. Wang, Z., Polygerinos, P., Overvelde, J. T. B., et al. (2017). Interaction forces of soft fiber reinforced bending actuators. *IEEE/ASME Transactions on Mechatronics, 22*(2), 717–727.
6. Peng, J. G., Li, D. K., Chen, D., et al. (2016). Comparison the characteristics of the finite element analysis software by the application in iron-based powder metallurgy industry. *Powder Metallurgy Technology, 34*(6), 461–466.
7. Lu, X. M., Xu, W., Li, X. (2015). Concepts and simulations of a soft robot mimicking human tongue. In *2015 6th international conference on automation, robotics and applications (ICARA)*, Queenstown, pp. 332–336. https://doi.org/10.1109/icara.2015.7081169.
8. Adriana, Č., Lana, V., Igor, K. (2017). Stress analysis of abdominal aortic aneurysm. *Zbornik radova Osmog susreta Hrvatskog društva za mehaniku/Penava, Davorin; Guljaš, Ivica; Bošnjak Klečina, Mirjana*, pp. 43–48.
9. Perruisseau-Carrier, A., Bahlouli, N., & Bierry, G. et al. (2017). Comparison between isotropic linear-elastic law and isotropic hyperelastic law in the finite element modeling of the brachial plexus. *Annales de Chirurgie Plastique Esthetique, 62*(6), 664–668.
10. Scharich, N., Schniter, B., Herbert, A. et al. (2017). Battery management system using Arduino. In *2017 IEEE Technology & Engineering Management Conference (TEMSCON)*, Santa Clara, CA, pp. 384–387. https://doi.org/10.1109/temscon.2017.7998405.

Chapter 5
Design and Simulation Analysis of Rigid-Flexible Hybrid Upper Limb Rehabilitation Mechanism

5.1 Introduction

With the change of population structure, there are more and more disabled and semi disabled, elderly people, while the proportion of the elderly in the total population is expanding. And at the same time, the number of patients with limb movement disorders due to stroke, spinal cord injury, cerebrovascular disease, and nervous system disease also increases year by year [1]. According to the research results of clinical medicine, the human brain can be remolded; then, the learning function can be reorganized through the repeated movement of the brain [2, 3]. Exercise therapy is beneficial to the recovery and improvement of motor function of hemiplegia patients and recover the patient's motor function in a short time [4–6]. Therefore, patients should carry out rehabilitation training in time through make use of synaptic plasticity to make up for the loss of nerve cell function, so as to restore limb function, which not only improves the quality of life of patients but also reduces family and social pressure [7, 8].

The development of new mechanical rehabilitation methods has inevitably led to the conceiving of specific recovery equipment. A flexible dynamic exoskeleton was jointly developed by Weiss Institute of Biological Engineering and Harvard John A. Paulson School of Engineering and Applied Sciences affiliated to Harvard University [9, 10]. The University of Salford designed the lower limb rehabilitation exoskeleton mainly for the elderly, the weak and the disabled [11]. Nerebot is a 3-DOF end traction rehabilitation robot designed by the University of Padua, Italy, which uses rope suspension control [12].

At present, there are two kinds of rehabilitation robots. One is the end actuator type, the other is the exoskeleton type [13].

The body of the end guided rehabilitation robot is placed on independent support, and the hand is tied on the end guide to achieving the purpose of rehabilitation training. The joint axis of this kind of robot doesn't need to coincide with the joints of the human body, so its structure is relatively simple. Early rehabilitation robots mainly focus on this direction. The end guided rehabilitation robot can't control the torque applied to each joint, so the reaction torque generated when the elbow joint moves directly act on the shoulder joint, and it will affect the patients with muscle weakness, In other words, it will directly accelerate the subluxation or even total dislocation of the shoulder joint. The exoskeleton rehabilitation robot and human limbs form a closed double chain motion mechanism. Through the adjustment of the link system, the patient's limbs can be assisted for rehabilitation training, which can achieve rich training actions. Because the exoskeleton rehabilitation robot has many advantages and can realize the requirements of active and passive in the process of rehabilitation training, the exoskeleton rehabilitation robot has gradually become the mainstream of rehabilitation robots [14], such as Armin II, 5 Rupert III, Caden. In order to ensure the safety and comfort of the human body, the shoulder joint movement center of the exoskeleton is coaxial with the human body movement center. At the same time, the elbow joint also needs relevant mechanisms to eliminate the mismatch problem. At present, many upper limb rehabilitation mechanisms are driven by motors, but the driving torque is very large, which makes the patients uncomfortable when doing rehabilitation exercise. In order to reduce their volume, many R&D teams can Wearing, using air pressure to drive, but the cylinder or pneumatic muscle is not easy to install, often make the rehabilitation mechanism look very complex. In this paper is proposed a new simple and low-cost pneumatic robotic mechanism for upper limb rehabilitation, Combine rigid cylinder with flexible pneumatic muscle. We employ a pneumatic actuator so that it is possible to obtain safety for the operation and the control of the force by the appropriate regulation of the pressures in the pneumatic cylinder chambers of the robot for rehabilitation. Finally, the feasibility and effectiveness of the proposed method are verified by simulation experiments.

5.2   Analysis of Upper Limb Joint Motion

The human body is composed of four major tissues. The most closely related part of the upper limb is mainly embodied in its muscle tissue and the bone tissue contained in its connective tissue. Only when the muscle and bone interact effectively, can the upper limb form a complex and multi-degree of freedom joint movement. According to the key points of motion, the upper limb of the human body can be divided into three more important joint parts: shoulder joint, elbow joint, and carpal joint.

Although the degree of freedom and amplitude of the three movements are different, there is a common point in addition. The movement of the three joints is realized by skeletal muscle. Each skeletal muscle is composed of two parts: muscle belly and tendon. The muscle belly is composed of contractile muscle fibers, while the tendon is composed of dense but not contractile connective tissue. Based on the basic shape of the physiological structure of skeletal muscle in different positions and the range of joint movement angle, the whole structure is carried out, including the radial wrist joint, the middle wrist joint, the carpometacarpal joint, the far end of the humerus and the radius, ulnar bone connection structure, the ball, and socket joint formed by the glenoid and the humeral head of the scapula, the imaginary transmission mechanism, the wearing way and the possibility of appearance movement obstruction and so on. Therefore, according to the similar physiological function characteristics of the executive parts that provide driving force, the mechanisms that meet the requirements are mainly selected as memory alloy, steering gear, motor, air bag, and cylinder. Through the detailed analysis of the motion principle of the above joints, the performance, principle, characteristics, basic information and organization difficulty of the expected several mechanisms are simulated and driven by the motion analysis, and finally confirmed as the combination scheme of the air bag and air cylinder.

The simple structure of each joint is analyzed as follows (Fig. 5.1).

Firstly, the shoulder joint, which connects the arm and trunk of the upper limb of the human body, can drive the arm to rotate around the shoulder joint for low pair. There are three degrees of freedom, which are: bending and extension in the vertical plane, abduction and adduction in the horizontal plane, internal rotation, and external rotation with the straight arm as the axis. Although each degree of freedom can move without the influence of other degrees of freedom, it is worth noting that the movement of any degree of freedom will affect the axis of other degrees of freedom.

Secondly, the elbow joint, as an elliptical joint, which connects the human body's big arm and forearm with the elbow joint, can form a rotation movement in different planes. It has two degrees of freedom, which are: flexion and extension in the vertical plane, internal rotation, and external rotation with the straight arm as the axis.

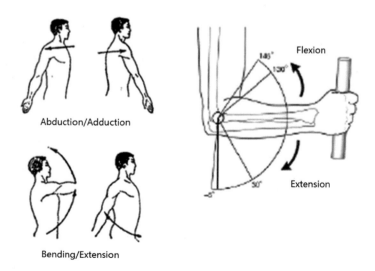

Fig. 5.1 Distribution of upper limb motion pairs

Through the detailed analysis of the motion principle of the above joints, among the executive parts that provide driving force similar to their physiological functions, the mechanisms that meet the requirements are mainly selected as follows: memory alloy, steering gear, motor, hydraulic, pneumatic muscle, and air cylinder. Through the simulation test and driving motion analysis of the performance, principle, characteristics, basic information, and organization difficulty of several mechanisms, it is finally confirmed that pneumatic muscle or cylinder is used to realize pneumatic hybrid drive.

Taking the process of the traditional rehabilitation training of the upper limb movement of stroke as an example, there are eight main processes: the range of joint activity training, joint loosening technology, drafting technology, muscle strength training, walking training, breathing training, balance and coordination training and aerobic training. Among them, the range of joint activity training includes three main training methods: the passive range of joint activity training, active-assisted joint activity training, and active movement. In the process of rehabilitation, the movement is required to be slow, soft, smooth, and repeated several times in a rhythmic manner, and maintained for several seconds after reaching the maximum range of activity as far as possible, to avoid impact sports and violence. Only looking at the motor drive mode and the steering gear drive mode, it can achieve a high transmission accuracy under a good control and modulation system, and compared with the hydraulic or pneumatic artificial muscle, cylinder, it does not need a large volume of liquid or another auxiliary air pump, so it is convenient to realize the overall movement and writing control of the mechanism. But in the needs of real clinical patients, weight and safety are the most important factors in the rehabilitation process. In the aspect of mechanism load on the human body, the weight of motor and steering gear on the upper body equipment is too heavy, and it is easy to produce unstable impact when starting and stopping, so it cannot fully meet the rehabilitation needs of patients. In the hydraulic scheme, although the response speed and control accuracy of its drivers are not as good as that of the motor and the steering gear, they are also relatively good. However, it does not meet safety requirements. For example, the transmission medium of the hydraulic drive is hydraulic oil. If the hydraulic oil leaks, it will cause harm to the user. Therefore, the rehabilitation of the exoskeleton program is left with the most suitable pneumatic muscle or cylinder two different drivers, which cooperate with each other to achieve a pneumatic hybrid drive.

This kind of pneumatic hybrid drive scheme not only has the comparative advantages between the above drivers but also has good embodiment in other aspects, such as miniaturization, lightweight, flexibility, and so on. The whole structure uses multiple air valves to control different air channels so that the position and speed of the joints to be recovered can be precisely controlled, and the training intensity can be adjusted. In addition, in some details, such as noise, it can also meet the needs of low working noise in the hospital environment.

5.3 Methodology

5.3.1 Cylinder Performance Analysis

The cylinder is a cylindrical metal part that guides the piston to move in a linear reciprocating motion. Air is transformed into mechanical energy by expansion in the engine cylinder; gas is compressed by piston in the compressor cylinder to increase the pressure. Compared with the electric actuator, the cylinder can work reliably under severe conditions, and the operation is simple, basically maintenance-free. The cylinder is good at reciprocating linear motion, especially suitable for the most transfer requirements in industrial automation linear handling of the workpiece. Moreover, only adjusting the one-way throttle valve installed on both sides of the cylinder can simply achieve stable speed control, which has become the biggest feature and advantage of the cylinder drive system. Therefore, for users who do not have the requirements of multi-point positioning, the vast majority prefer to use the cylinder from the perspective of ease of use. However, because of its large volume and small output load, much medical rehabilitation equipment will not use the cylinder as the driver (Fig. 5.2).

5.3.2 Tendon Performance Analysis

In the simplest case, the pneumatic tendon acts as a single-acting actuator, acting on a mechanical spring or load. Under the condition of pneumatic tendon expansion and no pressure, mechanical spring pretensions the pneumatic tendon in normal position. Ideal pretension: 0.5% of rated length. As far as the technical properties of pneumatic tendons are concerned, The following working state is ideal: when there is no pressure, the diaphragm will not be compressed. When the air intake is pressurized, the pneumatic tendon is pretensions to produce the maximum force, with the best dynamic characteristics and little air consumption. When the shrinkage is less than 9%, it can provide the most effective working range. The smaller the

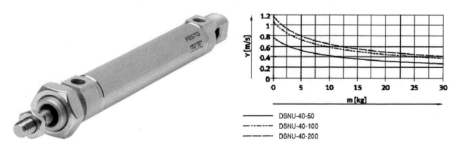

Fig. 5.2 Cylinder features

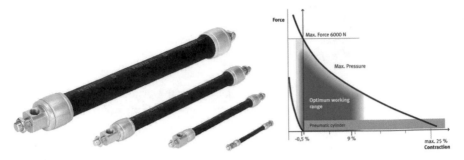

Fig. 5.3 Characteristics of the pneumatic tendon

contraction rate of the pneumatic tendon is, the higher its working efficiency is. When the external force changes, the working characteristics of the pneumatic tendon are like that of the spring: it follows the direction of the force. For pneumatic tendons, "pneumatic spring" and spring stiffness may change. When the pneumatic tendon is used as pneumatic spring, the pressure or volume is constant. This results in different spring characteristics, making the spring effect perfectly match the application. It can be concluded from the analysis that pneumatic tendon is widely used in medical rehabilitation because of its large output torque and soft strength, but its output displacement is small, the rehabilitation joint is limited, and it is mostly used in fixed driving rehabilitation institutions (Fig. 5.3).

5.4 Design and Simulation of Rehabilitation Mechanism

Through the analysis of the structure and movement form of each joint of the upper limb, through the study of the characteristics of the air cylinder and pneumatic tendon, this paper designs and studies the rigid-flexible hybrid mechanism of the shoulder joint and elbow joint of the upper limb, which is used for the research of upper limb rehabilitation. According to the structure and driving characteristics of the air cylinder and pneumatic tendon, the elbow joint is driven by an air cylinder, while the shoulder joint is driven by the pneumatic tendon. On the premise of ensuring the patients reach the standard of rehabilitation, comfort and safety, the wearable device can be realized by minimizing the volume and weight of the device.

The mechanism is divided into three parts: flexible spine, flexible shoulder and flexible shoulder (Fig. 5.4).

The flexible spine includes: pneumatic tendon, bionic spine of human body, inner line of relative displacement of monofilament nylon, outer line of relative displacement of monofilament nylon, pneumatic tendon adjusting movable pulley and pneumatic tendon adjusting fixed pulley; pneumatic tendon is distributed on both sides of the bionic spine of human body and connected with the pneumatic power source

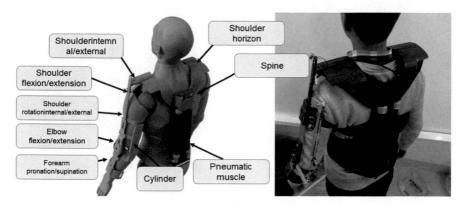

Fig. 5.4 Upper limb rehabilitation institution

through the switch solenoid valve; the top of pneumatic tendon is provided with pneumatic tendon adjusting movable pulley. The inner end of the relative displacement of monofilament nylon is fixed on the upper end of the pneumatic tendon adjusting pulley, after bypassing the pneumatic tendon adjusting fixed pulley, then bypassing the lower end of the pneumatic tendon adjusting pulley, and passing through the baffle to enter the outer line of the relative displacement of monofilament nylon.

The flexible shoulder includes: a shoulder fixing reinforcement and a shoulder fixing connector; two ends of the shoulder fixing reinforcement and the shoulder fixing connector are respectively fixed on the bionic spine, and the bionic upper limb; the outer line of the relative displacement of monofilament nylon is fixed with the shoulder fixing reinforcement, and the inner line of the relative displacement of monofilament nylon is connected with the big arm binding rod through the shoulder fixing reinforcement. The inner line of relative displacement of monofilament nylon pulls the big arm fitting rod to make its angle range with the shoulder fixed reinforcement 90°–0°.

The flexible shoulder includes: air cylinder actuator, big arm fitting rod, small arm fitting rod and rotating shaft; air cylinder actuator is connected with airpower source; air cylinder actuator is connected with small arm fitting rod and big arm fitting rod respectively through rotating pair; big arm fitting rod and small arm fitting rod are connected through rotating pair; after the gas enters the air cylinder actuator, the inner cylinder extends to drive the small arm to swing.

The driving structure of the whole elbow is a connecting rod mechanism as shown in the left side of Fig. 5.5. The connecting rod is divided into five parts: a, b, c, d, and e. Among them, a, b, c, e are fixed length rods, dare to be calculated length rods, in which aae, bc joints are fixed 90° joints, the rested, dc, ab joints are variable angle joints. We assume that the length of each segment of a, b, c, d, e is r_a, r_b, r_c, r_d, r_e, the movement angle of the forearm is θ, and the increasing length of the indefinite length rod d is f because the increasing length of the d rod f is transformed into the rotating movement of the rod around the ab joint and the rotating movement of the

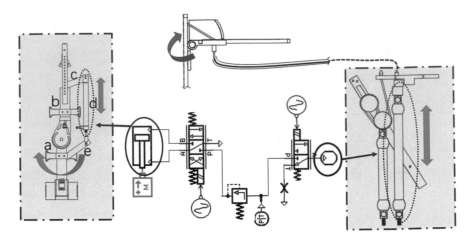

Fig. 5.5 Schematic diagram of rigid-flexible hybrid mechanism

e rod around the ae joint. Since the a and e are fixed angle links, the a-bar and E-bar can be simplified as the movement of the link with the length of $\sqrt{a^2 + e^2}$.

So, the circumference length of the rotating motion is:

$$2\pi \sqrt{r_a^2 + r_e^2} \tag{5.1}$$

Since the rotation angle is θ, the rotation motion length is:

$$\frac{2\pi\theta\sqrt{r_a^2+r_e^2}}{360} \tag{5.2}$$

So, the increased length:

$$f = \frac{2\pi\theta\sqrt{r_a^2+r_e^2}}{360} \tag{5.3}$$

So, the length of the indeterminate rod r_d:

$$r_d = \frac{2\pi\theta\sqrt{r_a^2+r_e^2}}{360} \tag{5.4}$$

We use pneumatic tendons and air cylinders with air pressure as the driving force to drive the exoskeleton to drive the patient's upper limb to move back and forth. Before making the real object, we need to establish a virtual prototype of the whole motion system, and carry out the virtual prototype simulation experiment of Kinematics/Dynamics in RecurDyn, in order to judge the effectiveness and rationality of the rehabilitation training of the exoskeleton to drive the patient's upper limb, we choose one of the most common physical rehabilitation methods for stroke patients to write the kinematic equation into the virtual prototype simulation platform for

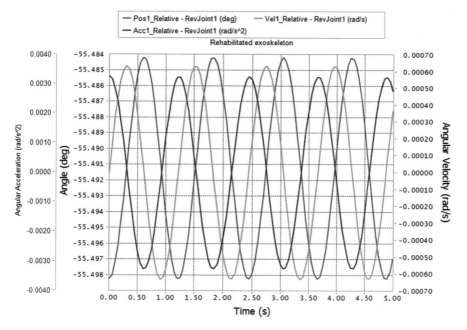

Fig. 5.6 Kinematics simulation results of shoulder

kinematic/dynamic simulation. Because we use the pneumatic tendon to pull the steel wire, it is difficult to control the movement of the upper arm of the exoskeleton. Therefore, we specially compare and analyze the kinematic simulation data on the shoulder of the exoskeleton. From the movement of the upper arm, we can get the simulation of the virtual prototype, which is in line with the normal speed of arm lifting in real life (Fig. 5.6).

In order to study the kinematic characteristics of the whole exoskeleton motion system, we combined the pneumatic tendon and the cylinder to carry out a common Kinematic/Dynamic simulation, and obtained the required rotation moment of the forearm movement, that is, the product of the force generated by the pneumatic tendon traction and the rotation radius, as well as the angular velocity image of the cylinder driven forearm movement (Fig. 5.7).

Through the simulation of the movement of the forearm, the required rotation moment can be provided for the experiment. As the rotation moment increases gradually, the angular velocity of the forearm also increases. In order to reduce the secondary injury to the body caused by the rehabilitation mechanism during the rehabilitation process, when the current arm moves to half of the end position, the rotation moment begins to decrease, and the angular velocity of the forearm reaches the maximum at this time. The rotation angle of the forearm presents a controllable linear relationship with time, which proves that the rehabilitation institution can effectively carry out rehabilitation training for patients according to the rehabilitation mode.

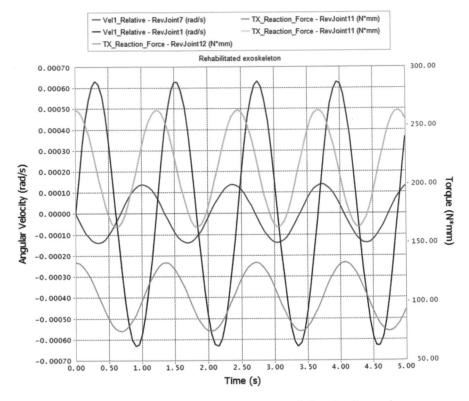

Fig. 5.7 Simulation results of upper arm rotation moment and elbow bending angle

5.5   Conclusion

Through the above simulation and analysis, the performance of the upper limb rehabilitation mechanism based on the rigid-flexible hybrid has been preliminarily verified. The design of the structure ensures the flexibility to meet the needs of complex daily activities, and the reasonable selection of air cylinder and pneumatic tendon ensures the flexible extension of each joint. The whole exoskeleton is light and small, with excellent durability and portability. The integrated driver circuit and power module can also be easily arranged on the back. The way of pneumatic drive also ensures the patient's safe rehabilitation training.

Funding   The study was supported by Key projects of national key research and development plan (2017YFF0207400): Research on key technologies and important standards of health services and remote health monitoring for the elderly and the disabled.

References

1. Chen, X. (2008). Data from the second national sample survey of disabled persons. *Chinese Journal of Reproductive Health, 19*(2), 68.
2. Xie, Z. C., Xu, G. L., & Liu, X. F. (2009). Research progress of early rehabilitation after stroke. *Chinese Journal of Rehabilitation Theory and Practice, 155*(10), 908–912.
3. Woldag, H., & Hummelsheim, H. (2002). Evidence-based physiotherapeutic concepts for improving arm and hand function in stroke patients: A review. *Journal of Neurology, 249*(5), 518–528.
4. Wang, Y. B., Ji, L. H., & Huang, J. Y. (2005). Development of a neural rehabilitation robot. *Mechanical Science and Technology, 24*(2), 139–141.
5. Li, J., Wang, X. W., & Yang, R. F. (2003). Effect of passive training on motor function of stroke patients in recovery period. *Chinese Clinical Rehabilitation, 7*(16), 2380.
6. Nef, T., & Riener, R. (2005). ARM in-design of a novel arm rehabilita-rehabilitation Robot. In *9th International Conference on Rehabilitation Robotics. ICORR 2005* (pp. 57–60). Chicago, IL.
7. Zhao, H. Y., & Zhang, M. Q. (2020). Study on debilitating status and influencing factors of elderly stroke patients with hemiplegia. *Health Vocational Education, 38*(1), 129–132.
8. Jiang, Q. H., Zhang, H. M., Wang, M., et al. (2019). A study on the status quo and influencing factors of self transcendence in elderly stroke patients. *China Journal of Practical Neurology, 20*, 1–8.
9. Asbeck, A. T., Dyer, R. J., Larussion, S. F., et al. (2013). Biologically-inspired soft exosuit. In *2013 IEEE 13th International Conference on Rehabilitation Robotics (ICORR)* (pp. 1–8). Seattle, WA. https://doi.org/10.1109/icorr.2013.6650455.
10. Awad, L. N., Bae, J., O'Donnell, K., et al. (2017). A soft robotic exosuit improves walking in patients after stroke. *Science Translational Medicine 9*(400), eaai9084.
11. AL-Fahaam, H., Davis, S., & Nefti-Meziani, S. (2016). Wrist rehabilitation exoskeleton robot based on pneumatic soft actuators. In *2016 International Conference for Students on Applied Engineering (ICSAE)* (pp. 491–496). Newcastle upon Tyne. https://doi.org/10.1109/icsae.2016.7810241.
12. Masiero, S., Armani, M., & Rosati, G. (2011). Upper-limb robot-assisted therapy in rehabilitation of acute stroke patients: Focused review and results of new randomized controlled trial. *Journal of Rehabilitation Research and Development, 48*(4), 355–366.
13. Lum, P. S., Burgar, C. G., Shor, P. C., et al. (2002). Robot-assisted movement training compared with conventional therapy techniques for the rehabilitation of upper-limb motor function after stroke. *Archives of Physical Medicine and Rehabilitation, 83*(7), 952–959.
14. Chen, W. H., Li, Z. Y., Cui, X., et al. (2019). Mechanical design and kinematic modeling of a cable-driven arm exoskeleton incorporating inaccurate human limb anthropomorphic parameters. *Sensors, 19*, 4461. https://doi.org/10.3390/s19204461.

Chapter 6
Classification and Treatment System for Facial Acne Vulgaris Based on Image Recognition

6.1 Introduction

Facial acne vulgaris is an inflammatory skin condition that occurs in the hair follicles and sebaceous glands [1]. Non-inflammatory skin lesions are characterized by open acne (blackheads) and closed acne (whiteheads) If patients are not treated promptly, they may cause scars, underlying mental disorders, and even depression [2]. Acne affects 9.4% of the world's population and is the eighth most prevalent disease in the world, usually occurring between the ages of 7–46, with the prevalence rate of acne in adolescence as high as 86.9% [3]. Boys are most frequently affected, especially in severe cases. 54.6% of the patients were treated by hospitals, beauty salons and skin care products [4]. For the treatment of acne vulgaris, dermatologists graded the patients through experience, but there were few experienced doctors, so a device was needed to quickly grade the acne for better treatment [5].

So far, in order to solve this problem, Fartash Vasefi [6] and Nicholas MacKinnon [7] have come up with a mobile medical application that spatially calibrates captured images based on reference points such as the location of the iris of the eye. Face recognition algorithm is used to recognize the features of face images, normalize the images, and define the facial area (ROI) of interest for acne evaluation [8]. Barbin [9] designed a facial fluid analysis using luminescence visualization system prototype by using optical imaging and fluorescence imaging system. Due to the different facial fluid area and acne area, they exclude other facial structure and leave with acne area mapping facial structure, and then judge the severity of the acne through the analysis of this facial structure [10].

The research content of this paper is mainly to use image recognition to detect and identify facial acne vulgaris, so that carry out graded diagnosis and treatment. Therefore, this paper designed a face acne vulgaris grading system based on image recognition. According to the shape and size of facial acne, we used the methods of area detection, circular contour detection and corner detection to detect the pre-processed facial acne image, so that determine the severity of acne and grade it.

6.2 Methods

6.2.1 Design of the Detection System

For the detection system of facial acne vulgaris grading diagnosis and treatment, the overall design includes image acquisition module, image preprocessing module, image recognition module and other parts (Fig. 6.1).

This detection system firstly collects local pictures of facial acne, and then carries out gray processing, filtering processing and binarization processing to show the features of acne through binarization images. Then the corner detection was used to mark the location of acne, and the contour detection was used to detect the contour of acne at the acne markers. In addition, area detection was used to distinguish acne of different areas so that mark acne of different severity. Through different detection methods, different detection thresholds are set to obtain different grading results. Finally, the results of various tests are integrated to obtain the final graded test results.

This paper adopts Python programming language, and combines OpenCV image processing library for image acquisition and image processing. In addition, the camera's controller is programmed in Python.

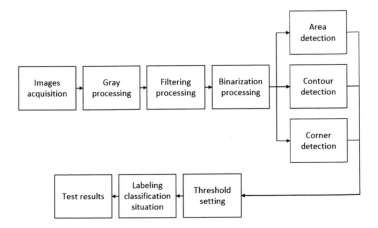

Fig. 6.1 Test block diagram

6.2.2 Design of Image Acquisition Device

This system designed an experimental device for collecting facial acne vulgaris, and mainly composed of a camera and LED light (Fig. 6.2). The camera is directly connected with the computer by USB data cable, and the computer controls the camera for image collection (Fig. 6.3). LED lamps are mainly used to supplement the light when collecting images.

Fig. 6.2 Detection device

Fig. 6.3 Installation location of equipment

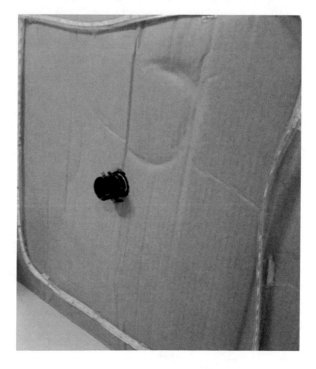

6.3 Analysis

6.3.1 Image Preprocessing of Facial Acne Vulgaris

The image of acne should be preprocessed before image recognition. In this paper, a camera was used to collect facial acne, and then the image was processed to obtain the gray image. Based on gray image, in order to eliminate the influence of facial villi, it is necessary to filter them. In this paper, gaussian filter is used to process the image.

Moreover, image enhancement was used to enhance the features of acne sites. Finally, Otsu method was used to treat the acne sites with binary value, and the pre-processed images were obtained. The pre-processed image shows the size and location of the acne (Fig. 6.4).

6.3.2 Analysis and Processing of Detection and Recognition Algorithms

As feature points in the image, small changes in any direction will cause great changes in the gray level. Therefore, in this paper, we will use Harris corner test to determine the location of acne.

Moreover, Sobel operator was used to obtain the edge information of the image, and then the contour of facial acne was drawn after contour detection. We subdivide it into two parts: the search for the image contour and the drawing of the found contour. Then we look for lesions in the images we have obtained, and we can distinguish the contour of the lesions from the normal skin color. Finally, through a series of threshold adjustment, we can get the contour of the lesion.

Above we draw the outline of the acne, here we need to our draw the outline of the area is calculated. We use the contour area calculation function to sum up the areas of all the contours. By setting the area size, if the patient is larger than a certain

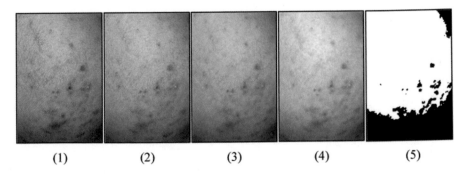

(1)   (2)   (3)   (4)   (5)

Fig. 6.4 Image preprocessing

Fig. 6.5 Corner detection
and contour detection

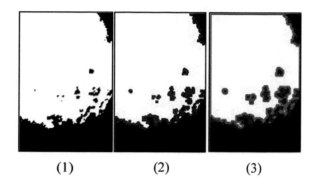

(1)   (2)   (3)

value, it is defined as severe. If the patient is within this value and outside a small value, it is defined as general; if the patient is less than this small value, it is defined as mild (Fig. 6.5).

6.4 Result

In this study, we used the above methods to identify facial acne of different severity, observe and analyze the recognize results. In this experiment, four facial acne patients with different severity were tested. The testing process included analyzing the severity of acne and marking the location and grade of acne. The results show that we can distinguish between the severity of facial acne vulgaris to a certain extent. The specific situation is to carry out area detection on the binary image of facial acne, and it can be concluded from area size analysis that the severity of facial acne vulgaris can be determined by different area sizes.

For the recognition reliability of facial acne vulgaris, according to the experiment, the acne area detected by contour is more than 90% consistent with the reality, and the failure is usually caused by the collection environment, image quality, and the defect of the recognition method proposed in this paper (Figs. 6.6, 6.7, 6.8 and 6.9).

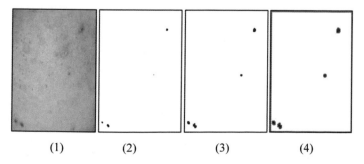

(1)   (2)   (3)   (4)

Fig. 6.6 Mild acne

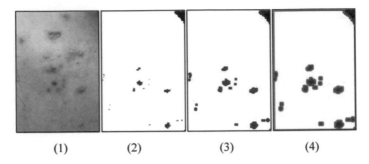

(1)   (2)   (3)   (4)

Fig. 6.7 Average acne

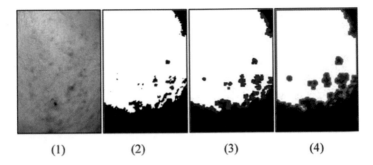

(1)   (2)   (3)   (4)

Fig. 6.8 Severe acne

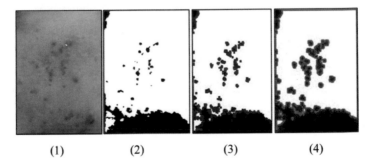

(1)   (2)   (3)   (4)

Fig. 6.9 Extreme acne

6.5 Conclusion

In this paper, we proposed a grading system for facial acne vulgaris based on image recognition. The system is designed to detect the presence and severity of acne vulgaris on the face. Through the methods mentioned in the above sections, medical personnel can know the severity of patients' facial acne, so that achieve the purpose of graded diagnosis and treatment. Therefore, this paper provides reference for grading diagnosis and treatment to some extent. However, the accuracy of this recognition technology still needs to be further studied, which to meet the higher requirements of practicality.

Funding The study was supported by Key projects of national key research and development plan (2017YFF0207400): Research on key technologies and important standards of health services and remote health monitoring for the elderly and the disabled.

References

1. Ahn, G. R., Kim, J. M., Park, S. J., et al. (2019). Selective sebaceous gland electrothermolysis using a single microneedle radiofrequency device for acne patients: A prospective randomized controlled study. *Lasers in Surgery and Medicine.* https://doi.org/10.1002/lsm.23152.
2. Chiricozzi, A., Faleri, S., Franceschini, C., et al. (2015). AISI: A new disease severity assessment tool for hidradenitis suppurativa. *Wounds-a Compendium of Clinical Research & Practice, 27*(10), 258–264.
3. Tan, J. K. L., & Bhate, K. (2015). A global perspective on the epidemiology of acne. *British Journal of Dermatology, 172*(Suppl 1), 3–12.
4. Xie, S. X., Zhang, Y. Q., Wang, L., et al. (2014). Epidemiological analysis on the treatment of acne patients in college students. *Chinese General Practice, 17*(19), 2265–2267.
5. Boen, M., & Jacob, C. (2019). A review and update of treatment options using the acne scar classification system. *Dermatologic Surgery, 45*(3), 411–422.
6. Balbin, J., Banhaw, R., Christian, M., et al. (2019). Caries lesion detection tool using near infrared image processing and decision tree learning. In *Conference: Fourth International Workshop on Pattern Recognition.* https://doi.org/10.1117/12.2540896.
7. Nicholas, M., Fartash, V., Nicholas, B., et al. (2016). Melanoma detection using smartphone and multimode hyperspectral imaging. *SPIE BiOS.* https://doi.org/10.1117/12.2222415.
8. Amini, M., Vasefi, F., Valdebran, M., et al. (2018). Automated facial acne assessment from smartphone images. In *Imaging, Manipulation, and Analysis of Biomolecules, Cells, and Tissues XVI.* https://doi.org/10.1117/12.2292506.
9. Balbin, J. R., Banhaw, R. L., Martin, C. R. O., et al. (2019). Caries lesion detection tool using near infrared image processing and decision tree learning. In *Caries Lesion Detection Tool Using Near Infrared Image Processing and Decision Tree Learning.* https://doi.org/10.1117/12.2540896.
10. Balbin, J. R., Cruz, J. C. D., Camba, C. O., et al. (2017). Facial fluid synthesis for assessment of acne vulgaris using luminescent visualization system through optical imaging and integration of fluorescent imaging system. In *Second International Workshop on Pattern Recognition. Society of Photo-Optical Instrumentation Engineers (SPIE) Conference Series.* https://doi.org/10.1117/12.2280829.

Chapter 7
Comparative Analysis Soft Kinematics of Hand Rehabilitation Robot Powered by Pneumatic Muscles

7.1 Introduction

Treatment of patients with injuries to muscles and nerves caused by stroke, surgery, car accident, neuromuscular disease or trauma through medical means, traditional rehabilitation training is often used by rehabilitation personnel such as occupational therapists or physical therapists. Provide the guidance and assistance of rehabilitation, help patients restore function, improve their ability to fully perform ability of daily living to restore normal life [1–3]. This has caused the efficiency of manual rehabilitation training to be affected by human factors, and for most patients, especially those in rural areas, it's difficult to pay for this artificial and expensive treatment [4]. The effectiveness and practicality of robot-assisted treatment have been shown in many studies to have a significant effect. Therefore, the design of a rehabilitation robot that can replace the treating physicians to a certain extent is one of the hot topics in recent years in the fields of mechanical design, electrical engineering, biomedical design, and computer reasoning and so on [5].

Although many exoskeleton hand robots have been created for different purposes, they all have their own special advantages and disadvantages [6]. Power-enhancing and rehabilitation wearable rehabilitation finger robots must have: intrinsically safe because they are in direct contact with humans; lightweight, easy to use and carry, and able to accommodate a wide range of human adjustments and calibrations [7]. Unfortunately, the current traditional driving brakes and some rigid, serial-linked rehabilitation robot opponents' injuries are secondary and are not very suitable for these requirements, and they are controllable, comfortable and safe for the rehabilitation process. It is also difficult to meet the requirements of safe rehabilitation [8]. Over the past two decades, assistive rehabilitation wearable hand exoskeleton robots have attracted many people's research and development interest. Hassanin Al-Fahaam et al. Have tried to develop new types of flexural aerodynamic muscles, and focused their research on their behavior on the mathematical modeling, however, no good tracking accuracy of the bending angle has been obtained [9].

The main purpose of this article is to introduce a novel wearable hand rehabilitation robot with 14 movable joints and the non-linear, time-varying and time-delay characteristics of the new pneumatic artificial muscle used at the finger joints used by the robot.

7.2 Experiment Setup

7.2.1 New Multi-chamber Pneumatic Bionic Artificial Muscle

This paper establishes the geometric model of the rehabilitation mechanical glove driver. Its size, shape and related shapes will affect the working range of the bending action of the driver. Based on the above factors, we establish a suitable geometric model of the rehabilitation hand, and we consider each finger joint Size and related dimensions, so we analyze the finger structure, size, installation and other factors to establish a geometric model of each joint (Figs. 7.1 and 7.2).

Fig. 7.1 Flexible muscle joints

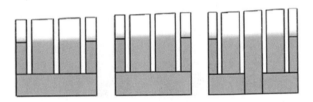

Fig. 7.2 Drive principle

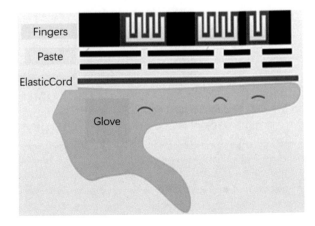

7.2.2 14-Joint Flexible Rehabilitation Gloves

Figures 7.3 and 7.4 show the wearing effect of our self-designed rehabilitation gloves. We made a 3D printed shell that conforms to the bending characteristics of each finger. There is a total of five fingers and 14 joints in the rehabilitation gloves, two

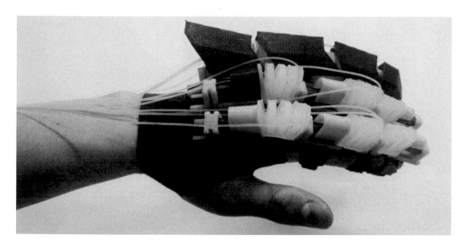

Fig. 7.3 Rehabilitation gloves

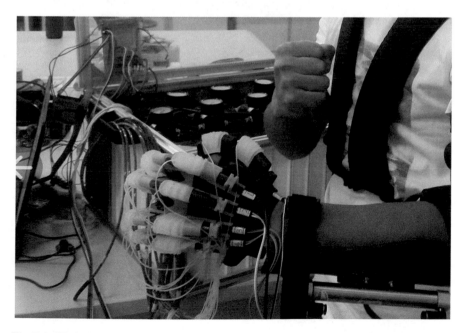

Fig. 7.4 Work situation

pneumatic bionic artificial muscles for the thumb and little finger, and three pneumatic bionic artificial muscles for the index finger, middle finger and ring finger. A pair of cylindrical motion pairs aligns each artificial muscle in reverse. The entire muscle group and passive sliders are fitted with Velcro and gloves conforming to human hand engineering characteristics for easy patient wear.

7.3   Aerodynamic Muscle Properties

7.3.1   Simulation Analysis and Experimental Verification

The reason why the silicone airbag expands is that we inflate the inside of the silicone airbag and cause internal pressure inside it [10]. So now, we want to simulate the expansion of the airbag and the analysis of the corresponding stress results. Set the analysis step and variable output, apply load and boundary conditions. The analysis step is mainly responsible for the analysis process. We can set the required incremental analysis step and the time required for this step. The variable output mainly represents the setting output result, including Stress, strain, etc. Apply load and external force to contact. The boundary condition is to set the degree of freedom of movement of the component [11]. After the above work is completed, we can enter the post-processing submission calculation. During the experiment, we determined to use the uniaxial tensile experiment method to determine the model parameters. The material model is often used when the finite element software is used to build the finite element model Is the two-parameter Mooney Rivlin equation, σ is the engineering stress value when the stretch ratio is γ. It can be seen that relationship between them is a straight line, that is, the slope C10 intercept is C01, where D1 represents the incompressibility, and our silicone is a very compressible material D1 = 0.

$$\frac{\sigma}{2(\lambda - 1/\lambda^2)} = C_{10} + C_{01}/\lambda \tag{7.1}$$

Arruda-boyce is a theoretical and practical method for considering rubber constitutive based on thermodynamics. The strain energy density based on Arruda-Boyce is defined as follows:

$$u = \mu \sum_{i=1}^{5} \frac{c_i}{\lambda_m^{2i-2}} (\overline{I_1^i} - 3^i) + \frac{1}{D}\left(\frac{J^2 - 1}{2} - \ln J\right) \tag{7.2}$$

Among them

$$C1 = 1/2 \quad C2 = \frac{1}{20} \quad C3 = \frac{11}{1015} \quad C4 = \frac{19}{7050} \quad C5 = \frac{519}{673750}$$

From the above cloud image results, we can see the expansion of the airbag. Based on the theory of the fourth strength of the Von Mise, the stress of the relevant parts is analyzed. The Fig. 7.5 is mainly reflected between the two adjacent airbags. There is stress concentration here. The structure of the airbag was modified. The simulation results of the airbag to verify the structure of the airbag almost matched the actual situation.

From the analysis of the results (Fig. 7.6), it can be seen that the results mainly show the magnitude of the stress at the maximum force point. As the air is continuously inflated, the internal pressure is continuously increased, and the stress at the maximum force point is continuously increased. This point is the danger point we

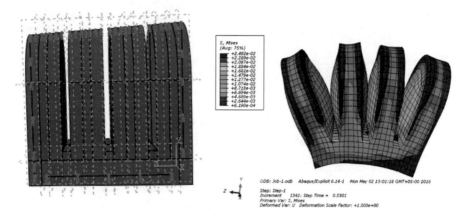

Fig. 7.5 Finite element simulation analysis

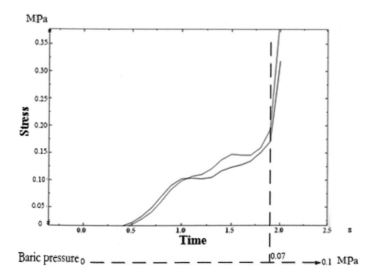

Fig. 7.6 Stress results

Fig. 7.7 In-kind presentation

have foreseen. We should try to avoid it. In addition, the actual situation is very close to our simulation results. With the continuous increase of pressure, when the pressure reaches 0.07 Mpa, the stress at the dangerous nodes increases rapidly, causing the airbag to break.

After obtaining the relevant data, this paper uses the calculation of the Mooney-Rivline rubber constitutive model to obtain C_{01}, C_{10}, and D_1, respectively. This is the basic parameter and the main parameter of our silica gel constitutive. Through the above simulation process and previous experiments, we can conclude that it is feasible to use silicone as a hand joint driver. In addition, the simulation results agree with the experiments. The following is an airbag model that we have improved by combining actual and simulation (Fig. 7.7).

In the new model, we have adaptively improved the size and details of the pneumatic artificial muscle. After the improvement, the ultimate pressure of the pneumatic artificial muscle was increased from 0.07 to 0.17 MPa.

7.4 Performance Analysis of Rehabilitation Fingers

Pneumatic artificial muscles are used as the driver of rehabilitation fingers, and a complete rehabilitation finger is added to the designed skeleton of rehabilitation fingers. Need to do kinematics simulation before making the real thing. The rehabilitation finger assembly diagram is as follows (Fig. 7.8).

Pneumatic rehabilitation finger can perform normal movement in kinematics simulation, and will enter the physical production. In this article, pneumatic muscles are assembled into a rehabilitation hand, and a complete rehabilitation finger is assembled to perform a mechanical test. Because the precise experimental equipment is too expensive, this article uses a mineral water bottle to add water as a load. Each time, a predetermined amount of water is added to the bottle for 50 g of water to perform

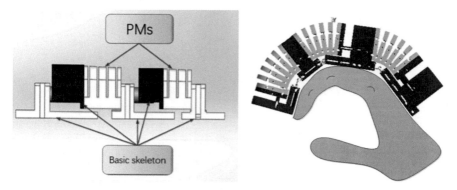

Fig. 7.8 3D, simulation diagram of rehabilitation finger

a rehabilitation finger strength test. At present, rehabilitation fingers can easily drag a load of 200 g (Fig. 7.9).

The following is the simulation result of the built pneumatic circuit. After the simulation calculation, the inflation amount of each airbag is simulated. Because the first joint has three airbags, three cylinders are used to simulate the inflation of the airbag. Each cylinder corresponds to a silicone airbag. The total distance at this time is 15 mm, and the amount of expansion on the airbag can be calculated from the above relationship (Fig. 7.10).

Fig. 7.9 Rehabilitation finger strength test

Fig. 7.10 Aerodynamic
simulation results

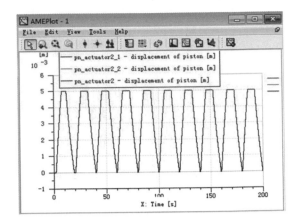

7.5 Conclusion and Future Works

With the gradual acceleration of the social rhythm, the importance of rehabilitation equipment has become increasingly prominent [12]. Rehabilitation with exoskeleton has become a gradually popular research direction in the field of rehabilitation equipment. However, traditional rigid exoskeleton is often driven by motors, steel wires rope, etc. There are great shortcomings in performance, rehabilitation safety, and patient comfort [13]. This paper introduces a new wearable rigid-flexible hybrid hand rehabilitation robot with multiple fingers and multiple joints. We analyze various performances of aerodynamic bionic artificial muscles designed and configured by us and verifies the correctness of the theoretical model through experiments. The wearable rigid-flexible hybrid hand rehabilitation robot with 14 movable joints has the characteristics of non-linear, time-varying and time-delay of pneumatic artificial muscles. The main components are passive sliders and pneumatic artificial muscles, which are firmly connected. When the muscles are inflated, the pneumatic muscles will gradually expand non-linearly with the pressure, and the passive slider will receive the axial force generated by the edges of the pneumatic muscles, thereby performing a cylindrical pair of movements. The combination of the two drives the fingers to achieve the effect of bending [14].

Funding The study was supported by Key projects of national key research and development plan (2017YFF0207400): Research on key technologies and important standards of health services and remote health monitoring for the elderly and the disabled.

References

1. Tiziani, L. O., & Hammond, F. L. (2020). Optical sensor-embedded pneumatic artificial muscle for position and force estimation. *Soft Robotics*. https://doi.org/10.1089/soro.2019.0019.
2. Nuckols, K. (2019). Proof of Concept of soft robotic glove for hand rehabilitation in stroke survivors. *Archives of Physical Medicine and Rehabilitation, 100*(12), e195.
3. Lane, K., Chandler, E., Payne, D., et al. (2020). Stroke survivors' recommendations for the visual representation of movement analysis measures: A technical report. *Physiotherapy, 107,* 36–42.
4. Anam, K., Rosyadi, A. A., & Sujanarko, B. (2018). The design of a low-cost therapy robot for hand rehabilitation of a post-stroke patient. In *2018 International Conference on Computer Engineering, Network and Intelligent Multimedia (CENIM)*. https://doi.org/10.1109/cenim.2018.8710833.
5. Goffredo, M., Mazzoleni, S., Gison, A., et al. (2019). Kinematic parameters for tracking patient progress during upper limb robot-assisted rehabilitation: An observational study on subacute stroke subjects. *Applied Bionics and Biomechanics, 2,* 1–12.
6. Vu Philip, P., Chestek Cynthia, A., Nason Samuel, R., et al. (2020). The future of upper extremity rehabilitation robotics: Research and practice. *Muscle and Nerve, 61*(6), 708–718.
7. Huang, Y. H., Nam, C. Y., Li, W. M., et al. (2020). A comparison of the rehabilitation effectiveness of neuromuscular electrical stimulation robotic hand training and pure robotic hand training after stroke: A randomized controlled trial. *Biomedical Signal Processing and Control, 56,* 101723.
8. Irshaidat, M., Soufian, M., Al-Ibadi, A., et al. (2019). A novel elbow pneumatic muscle actuator for exoskeleton arm rehabilitation. In *2019 IEEE International Conference on Soft Robotics (RoboSoft)*. https://doi.org/10.1109/robosoft.2019.8722813.
9. Yap, H. K., Lim, J. H., Nasrallah, F., et al. (2015). A soft exoskeleton for hand assistive and rehabilitation application using pneumatic actuators with variable stiffness. In *2015 IEEE International Conference on Robotics and Automation (ICRA)* (pp. 4967–4972). Seattle, WA.
10. Pawar, M. V. (2018) Experimental modelling of pneumatic artificial muscle systems designing of prosthetic robotic arm. In *3rd International Conference for Convergence in Technology (I2CT)* (pp. 1–6). Pune.
11. Pillsbury, T. E., Guan, Q., Wereley, N. M. (2016). Comparison of contractile and extensile pneumatic artificial muscles. In *2016 IEEE International Conference on Advanced Intelligent Mechatronics (AIM)* (pp. 94–99). Banff, AB.
12. Sarakoglou, I., Brygo, A., Mazzanti, D., et al. HEXOTRAC: A highly under-actuated hand exoskeleton for finger tracking and force feedback. In *2016 IEEE/RSJ International Conference on Intelligent Robots and Systems (IROS)* (pp. 1033–1040). Daejeon.
13. Zhou, Y., Zhang, P., Xiao, K., et al. (2017) Research on a new structure of hand exoskeleton for rehabilitation usage. In *2017 4th International Conference on Information Science and Control Engineering (ICISCE)* (Vol. 1, pp. 1126–1130). IEEE Computer Society.
14. Haghshenas-Jaryani, M., Carrigan, W., Nothnagle, C., et al. (2016). Sensorized soft robotic glove for continuous passive motion therapy. In *2016 6th IEEE International Conference on Biomedical Robotics and Biomechatronics (BioRob)* (pp. 815–820). Singapore.

Printed in the United States
By Bookmasters